Progress in Nonlinear Differential Equations and Their Applications
Volume 25

Variational Methods for Discontinuous Structures

Applications to image segmentation, continuum mechanics, homogenization

Villa Olmo, Como, 8–10 September 1994

Raul Serapioni
Franco Tomarelli
Editors

1996

Birkhäuser
Basel · Boston · Berlin

Raul Serapioni
Dipartimento di Matematica
Politecnico di Milano
Pzza. Leonardo da Vinci, 32
I–20133 Milano
Italy
e-mail: rauser@ipmma1.mate.polimi.it

Franco Tomarelli
Dipartimento di Matematica
Politecnico di Milano
Pzza. Leonardo da Vinci, 32
I–20133 Milano
Italy
e-mail: fratom@ipmma3.mate.polimi.it

1991 Mathematics Subject Classification 49-06, 35-06

A CIP catalogue record for this book is available from the Library of Congress, Washington D.C., USA

Deutsche Bibliothek Cataloging-in-Publication Data
Variational methods for discontinuous structures : applications to image segmentation, continuum mechanics, homogenization ; Villa Olmo, Como, 8 – 10 September 1994 / Raul Serapioni ; Franco Tomarelli ed. – Basel ; Boston ; Berlin : Birkhäuser, 1996
(Progress in nonlinear differential equations and their applications ; Vol. 25)
ISBN-13: 978-3-0348-9959-8 e-ISBN-13: 978-3-0348-9244-5
DOI: 10.1007/978-3-0348-9244-5

NE: Serapioni, Raul [Hrsg.]; GT

Softcover reprint of the hardcover 1st edition 1996

Printed on acid-free paper produced of chlorine-free pulp. TCF ∞

9 8 7 6 5 4 3 2 1

Contents

Preface

This volume contains the Proceedings of the International Conference *Variational Methods for Discontinuous Structures* which was organized by the *Dipartimento di Matematica del Politecnico di Milano*, and which took place at Villa Olmo on Lake Como, on 8–10 September 1994.

About one hundred researchers in various fields (calculus of variations, geometric measure theory, computer science and material science) attended the Conference. We take here the opportunity to thank them all for their contribution to the scientific discussions and to the friendly atmosphere of the event.

The meeting was successful in focusing on advanced results and techniques in pure and applied mathematics, and the scientific level of the lectures was remarkable.

Some of the talks were devoted to differential or variational modelization of image segmentation, domain partitions, surface flow by mean curvature, homogenization of composite materials, phase transformations, crystals, metastability of martensite, interaction between fracture and damage, magnetostrictive materials. In most cases theoretical and numerical analysis of these models was provided.

In the other talks specific problems in calculus of variations were isolated and focused with the aim of understanding and solving the main technical issues originated by the above models: noncoercive functionals, integral functionals with exponential growth or with different upper and lower growth estimates, new capacity notions connected with homogenization of nonlinear elliptic operators in perforated domains, Γ convergence, energy functionals depending both on sets and functions, functions with special bounded variation, free gradient discontinuities.

This volume contains contributions by 17 of the 21 speakers invited to deliver lectures during the Conference. A complete list of all speakers and of the authors of contributions is attached. Most of them presented original results, while the others gave surveys of new fields which are evolving rapidly at present.

The meeting would not have been possible without the financial support of several institutions and research projects. Here we wish to mention

- Progetto di ricerca C.N.R., Problemi Variazionali Irregolari
- Progetto di ricerca 40% M.U.R.S.T, Analisi Nonlineare e Calcolo delle Variazioni
- Gruppo Nazionale per l'Analisi Funzionale e le sue Applicazioni, C.N.R.
- Comitato 01 (Scienze Matematiche), C.N.R.

Moreover, we wish to thank the *Centro di Cultura scientifica Alessandro Volta* for logistic support and for having arranged the Conference sessions to be held in the beautiful neoclassic Sala Ovale of Villa Olmo. We also thank personable and efficient secretarial staff: Ms.Falascone, Ms.Tansini and Ms.Troglio.

Finally, we would like to thank all the Lecturers, and we are especially grateful to those who have contributed to the Proceedings.

Milano, 20 July 1995

Raul Serapioni
Franco Tomarelli

Invited Lectures

LUIGI AMBROSIO (Università di Pisa)
Flow by mean curvature of surfaces of any codimension

GABRIELE ANZELLOTTI (Università di Trento)
Functions of bounded variation over nonsmooth manifolds and generalized curvatures

JOHN M. BALL (Heriot Watt University, Edinbourgh)
Local minimizers and phase transformations

ALEXIS BONNET (Université de Cergy-Pontoise)
On the regularity of the edge set of Mumford-Shah minimizers

GUY BOUCHITTÉ (Université de Toulon)
Singular perturbations with a nonlocal term

ITALO CAPUZZO DOLCETTA (Università La Sapienza, Roma)
Indefinite superlinear elliptic problems

BERNARD DACOROGNA (Ecole Politecnique, Lausanne)
Remarks on a numerical study of convexity, quasi-convexity and rank-one convexity

GIANNI DAL MASO (S.I.S.S.A., Trieste)
Asymptotic behaviour of Dirichlet problems for perforated domains

ENNIO DE GIORGI (Scuola Normale Superiore, Pisa)
Movimenti di partizioni

IRENE FONSECA (Carnegie Mellon University, Pittsburgh)
Variational techniques for problems in material science

FEDERICO GIROSI (M.I.T., Boston)
Statistical mechanics approach to image segmentation and reconstruction

RICHARD D. JAMES (University of Minnesota, Minneapolis)
Metastability of martensite

DAVID KINDERLEHRER (Carnegie Mellon University, Pittsburgh)
Magnetoelastic interactions

ROBERT V. KOHN (Courant Institute, New York)
The crystalline algorithm for computing motion by curvature

ANTONIO LEACI (Università di Lecce)
A second order model in image segmentation: Blake & Zisserman functional

PAOLO MARCELLINI (Università di Firenze)
General growth conditions and regularity

MARIO MIRANDA (Università di Trento)
Geodesic lines in metric spaces

JEAN-MICHEL MOREL (Université Paris Dauphine)
A Kanisza programme

UMBERTO MOSCO (Università La Sapienza, Roma)
Homogeneous fractal spaces

JAYANT SHAH (Northeastern University, Boston)
Uses of elliptic approximations in computer vision

ITALO TAMANINI (Università di Trento)
Optimal approximation by piecewise constant functions

Progress in Nonlinear Differential Equations
and Their Applications, Vol. 25

Movimenti di Partizioni

ENNIO DE GIORGI

Scuola Normale Superiore, Piazza dei Cavalieri 7, I-56126 Pisa (Italia)

ABSTRACT. In questa esposizione viene presentata una congettura riguardante i movimenti di partizioni che generalizzano i movimenti di frontiere secondo la curvatura media. Se tale congettura venisse verificata, costituirebbe un interessante teorema di unicità e stabilità per un insieme denso di dati iniziali e probabilmente fornirebbe un modello interessante per lo studio di molte questioni analoghe.

Cominciamo perciò col dare alcune definizioni preliminari. Indicheremo con $\mathbf{R}^n$ lo spazio euclideo n-dimensionale e per ogni numero reale positivo o nullo α ed ogni sottoinsieme E di $\mathbf{R}^n$ indicheremo con $\mathcal{H}^\alpha(E)$ la misura di Hausdorff α-dimensionale di E. È ben noto che per $\alpha = n$ la misura $\mathcal{H}^\alpha$ coincide con la misura esterna di Lebesgue. Possiamo allora, per ogni coppia h, n di interi positivi, definire l'insieme di partizioni ordinate $\mathcal{P}(h,n)$ nel modo seguente:

DEFINIZIONE 1 Diremo che $\mathcal{C}$ appartiene a $\mathcal{P}(h,n)$ se soddisfa le condizioni seguenti:

1. $\mathcal{C}$ è una $(h+1)$-pla di sottoinsiemi chiusi di $\mathbf{R}^n$ che denoteremo con $C_1, \ldots, C_h, C_{h+1}$;
2. per ogni compatto $K \subset \mathbf{R}^n$ risulta $\displaystyle\sum_{i=1}^{h+1} \mathcal{H}^n(C_i \cap K) = \mathcal{H}^n(K)$;
3. per $i = 1, \ldots, h$, C_i è un insieme limitato;
4. per ogni $i \leq h+1$ e per ogni $x \in \mathbf{R}^n$, $x \in C_i$ se e soltanto se, scelto comunque $\rho > 0$, risulta $\mathcal{H}^n(B_\rho(x) \cap C_i) > 0$.

Nella famiglia $\mathcal{P}(h,n)$ si possono introdurre diversi tipi di distanza. Una prima distanza, che indicheremo con il simbolo $\ell(\mathcal{A},\mathcal{B})$, corrisponde in sostanza alla norma in L^1 della differenza delle funzioni caratteristiche; precisamente, se $\mathcal{A} = (A_1, \ldots, A_h, A_{h+1})$, $\mathcal{B} = (B_1, \ldots, B_h, B_{h+1})$, porremo

$$\ell(\mathcal{A},\mathcal{B}) = \sum_{i=1}^{h+1} \mathcal{H}^n(A_i \triangle B_i),$$

ove $\triangle$ indica la differenza simmetrica. Un'altra distanza estesa, cioè a valori nella semiretta estesa $[0 + \infty]$, che corrisponde in sostanza alla norma L^∞ della differenza delle distanze con segno, si può definire nel modo seguente: per ogni insieme $E \subset \mathbf{R}^n$ ed ogni $x \in \mathbf{R}^n$ poniamo

$$W(x,E) = \operatorname{dist}(x,E) - \operatorname{dist}(x,\mathbf{R}^n \setminus E);$$

poniamo poi per ogni coppia di insiemi E, L contenuti in $\mathbf{R}^n$

$$\overline{\delta}(E,L) = \sup_{x \in \mathbf{R}^n} |W(x,E) - W(x,L)|;$$

quest'espressione ha un valore ben determinato in tutti i casi tranne quelli $E = L = \emptyset$ e $E = L = \mathbf{R}^n$, nei quali porremo per definizione $\overline{\delta}(E,L) = 0$. Passando dagli insiemi alle partizioni porremo, per $\mathcal{A}$, $\mathcal{B}$ appartenenti a $\mathcal{P}(h,n)$,

$$\delta(\mathcal{A},\mathcal{B}) = \sum_{i=1}^{h+1} \overline{\delta}(A_i, B_i).$$

Infine possiamo considerare la funzione non simmetrica definita per ogni coppia di sottoinsiemi E, L di $\mathbf{R}^n$ dalla formula

$$\overline{\sigma}(E,L) = \int_{E \triangle L} \operatorname{dist}(x, \partial E) d\mathcal{H}^n(x).$$

Notiamo che $\overline{\sigma}(E,L) = 0$ se $\mathcal{H}^n\big(E \triangle L\big) = 0$, mentre $\overline{\sigma}(E,L) = +\infty$ se $\partial E = \emptyset$ e $\mathcal{H}^n\big(E \triangle L\big) > 0$. Quando $\mathcal{A}$, $\mathcal{B}$ sono due partizioni appartenenti a $\mathcal{P}(h,n)$, porremo poi

$$\sigma(\mathcal{A},\mathcal{B}) = \sum_{i=1}^{h+1} \overline{\sigma}(A_i, B_i).$$

Possiamo ora dare una definizione ispirata alle idee di Almgren, Taylor, Wang (cfr. [2]).

Definizione 2 Dati un'applicazione $\mathcal{M} : [0,+\infty[\to \mathcal{P}(h,n)$, un funzionale F definito in $\mathcal{P}(h,n)$ ed un elemento $\mathcal{G} \in \mathcal{P}(h,n)$, diremo che $\mathcal{M}$ è un movimento minimizzante F di origine $\mathcal{G}$ associato ad ℓ, δ, σ se soddisfa le condizioni seguenti: esistono un'applicazione $\mathcal{L} : \mathbf{N}^2 \to \mathcal{P}(h,n)$ ed una successione di numeri reali positivi $\{\lambda_s\}_{s \in \mathbf{N}}$ tali che

$$\lim_{s \to +\infty} \lambda_s = +\infty;$$

$$\lim_{s \to +\infty} \delta\big(\mathcal{L}(s,0), \mathcal{G}\big) = 0;$$

$$\lim_{s \to +\infty} \ell\big(\mathcal{M}(t), \mathcal{L}(\lambda_s, [\lambda_s t])\big) = 0,$$

ove per ogni $z \in \mathbf{R}$, $[z]$ indica la parte intera di z, e vale infine per ogni s, $i \in \mathbf{N}$ la relazione ricorrente

$$F\big(\mathcal{L}(s,i+1)\big) + \lambda_s \sigma\big(\mathcal{L}(s,i), \mathcal{L}(s,i+1)\big) = \min_{\mathcal{A} \in \mathcal{P}(h,n)} \big\{F(\mathcal{A}) + \lambda_s \sigma\big(\mathcal{L}(s,i), \mathcal{A}\big)\big\}.$$

La nozione ora data di movimento minimizzante può essere rafforzata in quella di movimento *stabile* minimizzante mediante la seguente

DEFINIZIONE 3 Ferme restando le ipotesi della Definizione 2, chiameremo $\mathcal{M}$ movimento stabile minimizzante F di origine $\mathcal{G}$ associato ad ℓ, δ, σ se gode delle seguenti proprietà:

1. $\mathcal{M}$ è l'unico movimento minimizzante F di origine $\mathcal{G}$ associato ad ℓ, δ, σ;

2. per ogni applicazione $\mathcal{B} : [1, +\infty[\to \mathcal{P}(h, n)$ verificante la condizione

$$\lim_{\lambda \to +\infty} \delta\big(\mathcal{B}(\lambda), \mathcal{G}\big) = 0$$

esiste $\mathcal{E} : [1, +\infty[\times \mathbf{N} \to \mathcal{P}(h, n)$ tale che $\mathcal{E}(\lambda, 0) = \mathcal{B}(\lambda)$,

$$F\big(\mathcal{E}(\lambda, i+1)\big) + \lambda\sigma\big(\mathcal{E}(\lambda, i), \mathcal{E}(\lambda, i+1)\big) = \min_{\mathcal{A} \in \mathcal{P}(h,n)} \big\{F(\mathcal{A}) + \lambda\sigma\big(\mathcal{E}(\lambda, i), \mathcal{A}\big)\big\}$$

e infine

$$\lim_{\lambda \to +\infty} \ell\big(\mathcal{M}(t), \mathcal{E}(\lambda, [\lambda t])\big) = 0.$$

Possiamo ora enunciare la congettura 1.

CONGETTURA 1 Sia F il funzionale definito ponendo per ogni $\mathcal{A} = (A, \ldots, A_h, A_{h+1})$ appartenente a $\mathcal{P}(h, n)$

$$F(\mathcal{A}) = \sum_{i=1}^{h+1} \mathcal{H}^{n-1}(\partial A_i).$$

In queste ipotesi è denso in $\mathcal{P}(h, n)$ rispetto alla distanza estesa δ l'insieme delle partizioni $\mathcal{G}$ per cui esiste un movimento stabile minimizzante F di origine $\mathcal{G}$ associato a ℓ, δ, σ.

OSSERVAZIONE 1 Nelle ipotesi della Congettura 1 si potrebbero considerare altre proprietà dei movimenti stabili. Per esempio, dato un movimento stabile $\mathcal{M}$, si potrebbe chiedere se per ogni t positivo anche $\mathcal{M}(t)$ è origine di un movimento stabile $\mathcal{M}'$ soddisfacente la condizione $\mathcal{M}'(\tau) = \mathcal{M}(t+\tau)$ per ogni $\tau \geq 0$. In sostanza questo ci direbbe che i movimenti stabili godono della proprietà di semigruppo.

OSSERVAZIONE 2 Nella Definizione 3, abbiamo imposto una convergenza piú forte ai valori iniziali $\mathcal{E}(\lambda, 0)$, una convergenza piú debole ad $\mathcal{E}(\lambda, [\lambda t])$ per $t > 0$. Volendo introdurre una convergenza forte sui "grafici", si potrebbe forse dimostrare che se

$\mathcal{M}$ soddisfa le condizioni indicate nella Definizione 3 allora per molti funzionali F la relazione

$$\lim_{\lambda \to +\infty} \delta\big(\mathcal{C}(\lambda), \mathcal{C}(\infty)\big) = 0.$$

è verificata dalle partizioni $\mathcal{C}(\lambda) \in \mathcal{P}(h, n+1)$ definite nel modo seguente:

per $i = 1, \ldots, h, h+1$, $\lambda \in [1, +\infty[$, $C_i(\lambda)$ è la chiusura dell'insieme

$$\{(x,t) : t \in [0, +\infty[,\ x \in E_i(\lambda, [\lambda t])\},$$

mentre $C_i(\infty)$ è la chiusura dell'insieme

$$\{(x,t) : t \in [0, +\infty[,\ x \in M_i(t)\}.$$

OSSERVAZIONE 3 Nelle ipotesi della Congettura 1, è probabile che per ogni $\mathcal{G} \in \mathcal{P}(h, n)$ esistano uno o piú movimenti minimizzanti F di origine $\mathcal{G}$ associati a ℓ, δ, σ. È probabile che se esiste uno solo di tali movimenti esso sia anche movimento stabile secondo la Definizione 3; piú difficile sembra stabilire se, nel caso in cui esistono parecchi movimenti aventi la stessa origine, essi siano tutti limiti di movimenti stabili aventi origini in punti vicini.

OSSERVAZIONE 4 Sarebbe interessante studiare il grado di regolarità delle frontiere degli insiemi $\mathcal{C}(\infty)$ introdotti nell'Osservazione 2. È probabile che si tratti di ipersuperfici di $\mathbf{R}^{n+1}$ di classe C^∞ fuori di un insieme singolare di misura $\mathcal{H}^n$ nulla. È anche probabile che questo primo risultato di regolarità sia ulteriormente migliorabile.

OSSERVAZIONE 5 La dimostrazione della Congettura 1 sembra difficile anche nel caso $n = 2$; forse converrebbe cominciare con l'esame dei casi piú semplici, per esempio quello in cui gl'insiemi $G_1, \ldots, G_h$ siano convessi di classe C^2 fra loro disgiunti, in tal caso è assai verosimile che anche gl'insiemi $M_1(t), \ldots, M_h(t)$, quando non sono vuoti, sono convessi le cui frontiere di classe C^2 si muovono secondo la curvatura media. Un altro caso semplice ma interessante è quello in cui $h = n = 2$, p è un numero positivo maggiore o uguale ad 1 mentre

$$G_1 = \{(x,y) :\ x^2 + (y-1)^2 \leq p,\ y \geq 0\}$$

$$G_2 = \{(x,y) :\ x^2 + (y+1)^2 \leq p,\ y \leq 0\}\ .$$

OSSERVAZIONE 6 La considerazione delle tre funzioni ℓ, δ, σ può apparire una complicazione eccessiva; d'altra parte, non sembra facile trovare definizioni piú semplici del movimento di partizioni che diano luogo agli stessi movimenti almeno nei casi semplici indicati nell'Osservazione 5.

Bibliografia

[1] **F.J. Almgren:** *Existence and regularity almost everywhere of solutions to elliptic variational problems with constraints,* Mem. Am Math. Soc., **165** (1976).

[2] **F.J. Almgren, J. Taylor, L. Wang:** *Curvature driven flows: a variational approach,* SIAM J. Control Optim. **31** 1993, 387–437.

[3] **G. Congedo, I. Tamanini:** *On the existence of solution to a problem in multidimensional segmentation,* Ann. Inst. Henri Poincaré **8** (1991), 175–195.

[4] **E. De Giorgi, G. Congedo, I. Tamanini:** *Problemi di regolarità per un nuovo tipo di funzionale del calcolo delle variazioni,* Atti Accad. Naz. Lincei, s.8 **82** (1988), 673–678.

[5] **U. Massari, I. Tamanini:** *Regularity properties of optimal segmentations,* J. reine angew. Math. **420** (1991), 61–84.

[6] **U. Massari, I. Tamanini:** *On the finiteness of optimal partitions,* Ann. Univ. Ferrara, **39** (1994), in stampa.

[7] **I. Tamanini:** *Optimal approximation by piecewise constant functions,* questo volume.

Progress in Nonlinear Differential Equations
and Their Applications, Vol. 25

The Crystalline Algorithm for Computing Motion by Curvature

PEDRO MARTINS GIRÃO⋆ AND ROBERT V. KOHN*

⋆ Instituto Superior Tecnico, Av. Rovisco Pais, Lisboa, Portugal
* Courant Institute, New York University, New York, NY 10012

ABSTRACT. Motion by (weighted) mean curvature is a geometric evolution law for surfaces, representing steepest descent with respect to (anisotropic) surface energy. The crystalline algorithm is a numerical method for computing this motion. The main idea of this method is to solve the analogous evolution law for a crystalline surface energy which approximates the underlying smooth one. We have recently explored the nature of this method, demonstrating its convergence in some simple special cases. This paper summarizes our results.

1. Introduction

This is an expository discussion of the crystalline algorithm for computing motion by curvature. Our goal is to explain the nature of this numerical scheme, what is known about it, and what remains open. We focus mainly on our recent work [18] and [19]. However, we also discuss related work by other authors.

2. Motion by weighted mean curvature

We are interested in the motion of phase boundaries represented by curves in the plane, or surfaces in three-space.

The definition of *weighted mean curvature* is by now standard. Curvature is defined by the first variation of surface area, and weighted mean curvature is the analogous object for an anisotropic surface energy. To make this more explicit, we begin with a surface energy density f defined on unit vectors. It determines the surface energy functional

$$E = \int_C f(n)\, ds \ ,$$

defined for any sufficiently regular phase boundary C. Here n is the inward unit normal; ds denotes arc length for curves in the plane and Hausdorff measure for surfaces in three-space. Now consider a one-parameter family of interfaces C_λ, obtained by moving each point $x \in C$ distance $\lambda h(x)$ in the normal direction. The first variation of surface energy has the form

$$\left.\frac{d}{d\lambda}\right|_{\lambda=0} E(C_\lambda) = -\int_C \omega h\, ds$$

for some $\omega : C \to \mathbb{R}$. This function ω is the weighted mean curvature of C. When the surface energy is isotropic ($f \equiv 1$) it becomes the mean curvature in the usual geometric sense. (Under our sign conventions, the curvature is $1/r$ when C is a circle of radius r.)

Motion by weighted mean curvature is a dissipative geometrical evolution, amounting to steepest descent for surface energy. Its equation is

$$V = \omega\ , \tag{1}$$

where V is the normal velocity of the interface.

Our interest arises from applications to materials science, see e.g. [20, 37]. In truth these applications rarely involve something so simple as (1). A more typical law is

$$V = M(n)[\omega + g(x, n)]\ ,$$

where $M(n)$ represents the mobility of the interface and $g(x, n)$ is the driving force. In many situations diffusion is crucial; then the driving force g depends on a temperature or concentration variable, which in turn satisfies a diffusion equation. We shall nevertheless concentrate on (1), viewing it as a special but central example and a laboratory for the development of new techniques.

We have not yet imposed any structural hypotheses upon the surface energy density f. We are mainly interested in "nice" surface energies – smooth functions f for which the associated evolution equation (1) is parabolic and well-posed. There are several equivalent characterizations of the acceptable f's. One is that the Frank diagram (the polar plot of $1/f$) should be strictly convex. Another is that the function $\xi \to |\xi| f(\xi/|\xi|)$ should be convex, and strictly so except in the radial direction. For the motion of curves in the plane we can write $f = f(\theta)$ by taking $n = e^{i\theta}$; the equivalent condition is then $f(\theta) + f''(\theta) > 0$. See for example [6], [35] or the appendix of [19] for this basic material.

There are by now many numerical schemes for solving (1) and its generalizations. A comprehensive review is beyond the scope of this paper. We nevertheless mention several approaches, with selected references and apologies in advance to those whose work is omitted.

Front-tracking is the most classical and accurate method, especially for curves in the plane, but it has difficulty accommodating changes of topological type. The main idea is to represent the moving interface as a time-dependent, parameterized curve $s \to \phi(s, t)$. This leads to a nonlinear parabolic equation for ϕ. Current ideas on efficient solution techniques can be found in [21].

The **level-set method** is especially convenient for problems involving complex initial data or changes of topological type. The main idea is to represent the interface as the level set of a function $\{x : v(x, t) = 0\}$. The condition that the interface move by curvature is a degenerate parabolic equation $v_t/|\nabla v| = \operatorname{div}(\nabla v/|\nabla v|)$. First introduced as a numerical tool [26, 31], this approach has also become an essential analytical technique [12, 15]. Some convergence results are given in [39].

The **Allen-Cahn equation** is of interest partly because it makes contact with the phase–field approach to solidification. The interface is again a level set $\{x : u(x,t) = 0\}$, but the function $u(x,t)$ solves a different equation $u_t - \Delta u + \epsilon^{-2}(u^3 - u) = 0$. Motion by curvature is achieved in the limit $\epsilon \to 0$, not only for classical solutions but also for singular ones [9, 14, 32]. The Allen-Cahn approach appears awkward numerically, because it requires the solution of a very stiff system. Phase field models have, however, been implemented numerically with considerable success, see for example [10, 22, 38, 40, 41]. A different but related "double obstacle" approach to motion by curvature is discussed in [24], and convergence results are given in [25].

The **time-implicit method** solves a geometric variational method at each time step. The theory of flat curvature flows developed in [2] makes use of such an approximation. Numerically, a scheme of this type was used to simulate solidification in [3].

A **time-splitting method** was recently introduced in [23]. The main goal of that work was to deal with multiphase problems, however the method is also interesting in the two-phase context. The basic iteration has two parts: first one blurs the interface by diffusion, then one sharpens it again. Convergence results are given in [8, 13].

Our list omits the **crystalline algorithm**, since it is the focus of the rest of this paper.

3. Crystalline surface energies

We said we were mainly interested in "nice" surface energies, i.e. smooth f's with strictly convex Frank diagrams. But that is not the whole truth. We are also interested in "crystalline" surface energies, by which we mean those whose Frank diagrams are *convex polygons*[1]. (Remember, the Frank diagram is the polar plot of $1/f$, in other words the curve defined in polar coordinates by $r = 1/f(e^{i\theta})$.)

A crystalline energy determines a finite family of distinguished normal vectors — the vertices of the Frank diagram. It is natural to consider the restricted class of piecewise linear surfaces with just these normals. Such surfaces have no curvature in the conventional geometric sense (more precisely, the curvature is zero on each face and infinite at each vertex). But one can still define their weighted mean curvature *variationally*: taking the first variation of surface energy within this *restricted* class of surfaces associates to each face a well-defined (constant) weighted mean curvature. Angenent and Gurtin [6] and Taylor [33, 36] used this observation to define motion by weighted mean curvature in the crystalline case. If no new faces are created, then the equation $V = \omega$ amounts to a system of ODE's for the normal displacements of the initial faces.

[1]Our use of the term "crystalline" is not quite the standard one. Many authors would call a nonconvex surface energy crystalline if the *convexification* of its Frank diagram is a convex polygon. In this paper, we never consider surface energies with nonconvex Frank diagrams.

This idea works better in the plane than in three-space: for surfaces in $\mathbb{R}^3$ the resulting system of ODE's may cease to have a solution in finite time [36]. This shouldn't be too puzzling – it just means that the restricted class of surfaces was too small. According to Taylor, the motion must be continued as a varifold.

Even in the plane the situation is more subtle than it seems. We assumed above that no new faces are created. For problems of the type $V = \omega + g$ this hypothesis seems to be wrong. The shattering process discussed in [28, 29, 30] is specifically designed to detect the creation of new faces.

These difficulties don't arise, however, when solving $V = \omega$ for curves in the plane with crystalline initial data. This assertion is far from self-evident; it follows from the analysis summarized herein and also from the work of Taylor [36]. As a result, the use of a crystalline rather than smooth surface energy radically simplifies the problem of motion-by-curvature for plane curves, by replacing a PDE with a system of ODE's.

4. Crystalline approximation as a numerical method

Thus crystalline approximation presents itself as a physically natural numerical method for computing motion by curvature [27]. Suppose our goal is to compute the evolution of a smooth curve C, under the motion by weighted curvature associated with a convex surface energy f. The crystalline algorithm (i) approximates f with a crystalline surface energy $\tilde{f}$, then (ii) approximates C with a polygonal curve $\tilde{C}$ whose normals lie in the directions preferred by $\tilde{f}$, then finally (iii) solves the ODE's describing motion by weighted curvature for this crystalline problem.

What kind of method is this? Basically, a sort of *nonlinear Galerkin scheme.* A traditional Galerkin method for solving a nonlinear evolution equation starts with a basis $\{\phi_k(x)\}_{k=1}^{\infty}$. One seeks an approximate solution of the form $u(x,t) = \sum_{k=1}^{N} a_k(t)\phi_k(x)$ by solving a suitable ODE for the coefficients $a_k(t)$. In crystalline approximation the spatial discretization is different and more nonlinear: rather than restrict $u(\cdot,t)$ to a finite dimensional linear space, we restrict the normals of the moving curve instead.

Curiously, for closed convex curves in the plane crystalline approximation can also be viewed quite differently. When expressed in appropriate variables it becomes a sort of *finite difference scheme*, as we shall explain below.

5. Graphs in the plane

To examine this method in detail, we now consider the motion of a graph in the plane [19]. The unknown is a scalar-valued function $u(x,t)$ defined for $x \in [0,1]$ and $t > 0$. Its (anisotropic) surface energy can be expressed as

$$\int_0^1 W(u_x)\, dx$$

with W convex, and motion by weighted curvature is described by the quasilinear parabolic PDE

$$\frac{u_t}{(1+u_x^2)^{1/2}} = W''(u_x)u_{xx}\ . \tag{2}$$

We shall suppose that the ends are pinned, so the boundary condition is $u(0,t) = u(1,t) = 0$.

The crystalline algorithm considers only functions with special tangents. This means we permit only special values of u_x. The list of permitted u_x determines a piecewise linear approximation to W, whose graph has corners at the associated points $(u_x, W(u_x))$. This is our crystalline energy.

We now specify the discrete equation of motion. Let

$$m = \text{the maximum distance between permitted values of } u_x; \tag{3}$$

this is the "mesh size" of the crystalline approximation. The numerical solution $u^m(x,t)$ should be piecewise linear in x at each time t, with vertices at $0 = x_0 < x_1(t) < \ldots < x_{N-1}(t) < x_N = 1$. Its slopes may take only permitted values, with adjacent segments taking adjacent slopes. We shall refer to the segment between x_{i-1} and x_i as the ith face. Let $(u_x^m)_i$, $(u_t^m)_i$ be the (constant-in-space) values of u_x^m and u_t^m on the ith face. A calculation using only the continuity of u^m gives

$$\dot{x}_i = -\frac{(u_t^m)_{i+1} - (u_t^m)_i}{(u_x^m)_{i+1} - (u_x^m)_i} ,$$

so the motion of the curve is fully determined by specifying the values of $(u_t^m)_i$. They are determined, of course, by the crystalline analogue of (2). The first variation of crystalline energy turns out to be $-\Delta_i/L_i$ on the ith face, where $L_i = x_i - x_{i-1}$ is the length of the ith face,

$$\Delta_i = \frac{W[(u_x^m)_{i+1}] - W[(u_x^m)_i]}{(u_x^m)_{i+1} - (u_x^m)_i} - \frac{W[(u_x^m)_i] - W[(u_x^m)_{i-1}]}{(u_x^m)_i - (u_x^m)_{i-1}}$$

for $2 \le i \le N-1$, and $\Delta_1 = \Delta_N = 0$. So the discrete evolution is

$$(u_t^m)_i = \sqrt{1 + (u_x^m)_i^2} \cdot \frac{\Delta_i}{L_i} . \tag{4}$$

We must also discretize the initial data, i.e. $u^m(x,0)$ must be a crystalline approximation of $u(x,0)$. One convenient method starts by drawing those tangents to the graph of $u(x,0)$ which lie in permitted directions. They can be pieced together (modifying them a bit at the endpoints) to obtain $u^m(x,0)$.

Viewed this way, there is nothing intrinsically geometric about the crystalline approximation scheme. It is a nonlinear Galerkin scheme, suitable for solving any parabolic equation of the form $u_t = g(u_x)W''(u_x)u_{xx}$. It can also be used to solve the *linear* heat equation, $u_t = u_{xx}$: then (4) simplifies to

$$(u_t^m)_i = \frac{(u_x^m)_{i+1} - (u_x^m)_{i-1}}{2L_i} .$$

In deriving the crystalline motion, we assumed that neighboring faces of u^m have neighboring slopes u_x^m. One can show that this property is preserved by the flow, even as faces disappear [19, 36]. Therefore the crystalline motion is well-defined. (One must of course re-index at the times when faces disappear.)

Our paper [19] demonstrates the convergence of this scheme. The analysis is simplest and the result strongest for the linear heat equation. In that case

$$\frac{d}{dt}\int_0^1 |u_x - u_x^m|^2\,dx \leq -2\int_0^1 |u_t - u_t^m|^2\,dx + m\|u_{xxx}\|_\infty(t)\ , \tag{5}$$

where m is defined by (3). The proof resembles the standard one for a linear Galerkin method: it uses only the evolution equations, the boundary conditions, and Taylor expansion of u. We may suppose that the initial data satisfies

$$\max_x |u_x - u_x^m| \leq m \quad \text{at } t = 0,$$

for example by choosing it as indicated above. Then (5) leads immediately to the H^1 convergence result:

$$\int_0^1 |u_x - u_x^m|^2\,dx \leq m^2 + m\int_0^t \|u_{xxx}\|_\infty(\tau)\,d\tau \quad \text{for each } t \geq 0.$$

The situation is similar though more complicated for the nonlinear equation (4). One must use the Taylor expansion of W as well as that of u, and the resulting differential inequality is

$$\frac{d}{dt}\int_0^1 |u_x - u_x^m|^2\,dx \leq A(t)\int_0^1 |u_x - u_x^m|^2\,dx + B(t)m$$

rather than (5). Here $A(t)$ and $B(t)$ depend on the (smooth) solution $u(x,t)$ but they are independent of m. An application of Gronwall's inequality gives once again a convergence result of the form

$$\max_{0\leq t\leq T} \|u - u^m\|_{H^1([0,1])} \leq C_T m^{1/2}\ .$$

Our discussion has focussed on the homogeneous Dirichlet boundary condition $u(0,t) = u(1,t) = 0$, but the situation is similar for the homogeneous Neumann boundary condition $u_x(0,t) = u_x(1,t) = 0$.

6. Closed convex curves in the plane

We have shown that the crystalline approximation scheme is a sort of nonlinear Galerkin method. Curiously, however, when applied to closed convex curves and viewed in the proper variables it looks more like a finite difference scheme [18].

The motion of a convex curve in the plane has a convenient curvature–angle representation, in which its (weighted) curvature ω is viewed as a function of time and the angle θ between its normal and some fixed axis. This is very well known for isotropic motion by curvature; the more general, anisotropic case is treated for example in [6]. If f is the surface energy density represented as a function of θ, then the weighted curvature ω can be expressed in terms of the ordinary curvature κ as

$$\omega = (f + f'')\kappa\ ,$$

and motion by weighted curvature is expressed by the PDE

$$\omega_t = (f + f'')^{-1}(\omega^2 \omega_{\theta\theta} + \omega^3) \ . \tag{6}$$

The crystalline algorithm considers only curves with special tangents, i.e. convex polygons. The faces are restricted to certain directions, corresponding to a list of permitted angles $\{\theta_i\}_{i=1}^N$. In the following analysis we take these angles to be equally spaced:

$$\theta_i = i\Delta\theta \ , \quad \Delta\theta = 2\pi/N \ .$$

There is an associated crystalline energy. It agrees with f at the special angles θ_i, and its Frank diagram is the N-sided polygon determined by these vertices.

The discretized evolution is motion by weighted mean curvature for this crystalline energy. It says

$$V_i = \omega_i \ , \tag{7}$$

where V_i and ω_i are respectively the normal velocity and weighted curvature of the ith face (the face with angle $i\Delta\theta$). We must also discretize the initial data, replacing a smooth convex curve by a polygon with permitted faces. The method suggested earlier for graphs works well here too: it forms the initial polygon as a union of line segments tangent to the initial curve at the points with permitted normals.

With some work, (7) can be put in a form analogous to (6). We outline the main steps, referring to [18] for the details. Starting from the definition of ω as the first variation of surface energy one finds that

$$\omega_i = \left[f_i + \frac{f_{i+1} - 2f_i + f_{i-1}}{2(1 - \cos\Delta\theta)} \right] \kappa_i$$

where

$$\kappa_i = 2\tan\left(\frac{\Delta\theta}{2}\right)\frac{1}{L_i}$$

is the curvature of the ith face (defined by the first variation of isotropic surface energy), $f_i = f(i\Delta\theta)$, and L_i is the length of the ith face. Also, from kinematic considerations, a solution of (7) has

$$\dot{L}_i = 2\omega_i \cot\Delta\theta - \omega_{i-1}\csc\Delta\theta - \omega_{i+1}\csc\Delta\theta \ .$$

Using these formulas, one arrives with some calculation at the desired analogue of (6):

$$\dot{\omega}_i = \left[f_i + \frac{f_{i+1} - 2f_i + f_{i-1}}{2(1 - \cos\Delta\theta)} \right]^{-1} \left[\omega_i^2 \frac{\omega_{i+1} - 2\omega_i + \omega_{i-1}}{2(1 - \cos\Delta\theta)} + \omega_i^3 \right] \ . \tag{8}$$

This is very nearly a standard finite-difference approximation to (6)! The only difference is the use of $2(1 - \cos\Delta\theta)$ rather than $(\Delta\theta)^2$ in the denominator. These two expressions are the same to principal order as $\Delta\theta \to 0$, so the convergence of this scheme is hardly in doubt.

There is, however, some work to do. In deriving the equation of motion we assumed that each ω_i was well defined, i.e. that the evolving polygon makes use of each permitted angle. This can and must be proved. In fact, as the polygon shrinks its area $A_{\Delta\theta}(t)$ decreases at a constant rate

$$\dot{A}_{\Delta\theta} = -\sum_i \frac{2(1-\cos\Delta\theta)}{\sin\Delta\theta}\left[f_i + \frac{f_{i+1}-2f_i+f_{i-1}}{2(1-\cos\Delta\theta)}\right]$$

until the extinction time $T_{\Delta\theta}$ when $A_{\Delta\theta} = 0$, and no face disappears prior to $T_{\Delta\theta}$. This is the crystalline analogue of the theorem of Gage and Hamilton [17], that a smooth convex curve moving by curvature in the plane remains smooth and convex until its extinction time $T = A(0)/\int_0^{2\pi}(f+f'')\,d\theta$. Interestingly, the proof that no face disappears before the extinction time uses the presence of $2(1-\cos\Delta\theta)$ rather than $(\Delta\theta)^2$ in (8).

We turn to the matter of convergence. Let $\omega(\theta,t)$ describe the smooth curve as it moves by weighted curvature, and let $\{\omega_i(t)\}$ be the crystalline solution described above with $\Delta\theta$ sufficiently small. Our method of choosing the initial polygon turns out to satisfy

$$|\omega(i\Delta\theta,0) - \omega_i(0)| \le C(\Delta\theta)^2\ .$$

Since the difference approximation (8) is second-order accurate, it is not hard to deduce

$$|\omega(i\Delta\theta,t) - \omega_i(t)| \le C_\tau(\Delta\theta)^2 \quad \text{for } t\le\tau,$$

for any τ less than the extinction time. A similar argument also shows

$$|\omega_\theta(i\Delta\theta,t) - \frac{\omega_{i+1}(t)-\omega_i(t)}{\sin\Delta\theta}| \le C_\tau\Delta\theta \quad \text{for } t\le\tau.$$

This gives convergence of the crystalline scheme in curvature–angle coordinates. However the natural question is the convergence of the curves themselves. This requires some extra work; the result proved in [18] is

$$D(C(t), P_{\Delta\theta}(t)) \le C_\tau(\Delta\theta)^2 \quad \text{for } t\le\tau,$$

where $C(t)$ is the evolving smooth curve, $P_{\Delta\theta}(t)$ is the crystalline approximation, and D is the Hausdorff metric.

7. Alternative viewpoints

We have carefully avoided discussing crystalline motion by curvature for non-crystalline initial data, i.e. for curves whose initial normals are not restricted to the vertices of the Frank diagram. We have no need to consider such curves, since a numerical scheme is entitled to use approximate initial data. In our treatment of graphs, the error due to approximating the initial data is small compared to the error due to approximating the PDE; for simple closed curves the two errors are of the same order.

Other authors have taken a different but equally valid viewpoint. Angenent and Gurtin [6] and Taylor [36] consider it important to discuss crystalline motion by curvature for non-crystalline initial data, because some real materials are described by crystalline surface energies. The treatments just cited are based on approximating noncrystalline data by oscillatory crystalline data, then passing to a suitable limit.

The crystalline evolution of graphs in the plane has also been considered by Fukui and Giga [16]. We summarize briefly their approach. Remembering (2), one is tempted to say that the crystalline evolution solves

$$u_t = \sqrt{1+u_x^2}W''(u_x)u_{xx}$$

with W piecewise linear. This equation doesn't make sense as written, but it is formally equivalent to

$$u_t = (\bar{W}'(u_x))_x \tag{9}$$

with $\bar{W}'(t) = \sqrt{1+t^2}W'(t) - \int_0^t(\sqrt{1+s^2})'W'(s)\,ds$. There is an extensive theory concerning divergence-form heat equations such as (9), including theorems of existence and well-posedness even when $\bar{W}$ is merely convex and piecewise C^1. This gives an alternative definition of crystalline evolution. Fukui and Giga show that for crystalline initial data, the solution obtained this way is identical to the one obtained using (4).

The work of Fukui and Giga has two important consequences. First, it applies even to noncrystalline initial data, giving a definition of such motion which is clearly well-posed and independent of any specific method for approximating the initial data. Second, their work gives another proof of convergence of the crystalline approximation scheme for graphs, since the solution depends continuously on the surface energy as well as on the data. However this argument does not give a convergence rate.

We have thus far discussed only curves in the plane, using methods that are intrinsically one-dimensional. A much more general approach has been developed by Almgren, Taylor, and Wang [2]. They prove the existence of a "flat curvature flow" – essentially, a weak solution of motion by weighted curvature – for a broad class of surface energies, including crystalline ones. Their approach is something like an implicit numerical method: it starts with a discrete-in-time solution, defined by solving an appropriate variational problem at each time step; then estimates are proved yielding the existence of a continuous-time limit.

This approach gives existence but not uniqueness. However, that is not the fault of the method: flow by weighted mean curvature is not necessarily unique. There are examples of smooth hypersurfaces, even in $\mathbb{R}^3$, whose flow by (isotropic) curvature develops a "fattening singularity" and ceases to be unique thereafter [5, 7]. For graphs in the plane and convex closed curves the flat curvature flow is unique, since it agrees with the classical solution so long as one exists.

When applied to a crystalline energy, the flat curvature flow provides an alternative definition of motion by weighted curvature. But as in the work of Fukui and Giga, it has to be shown that this is the same as the evolution described by (4) or (7) when the initial data is crystalline. This fundamental consistency result is proved for curves in the plane in [1], provided that in the crystalline solution two adjacent faces do not vanish simultaneously.

The paper [2] does not discuss continuity with respect to perturbations of the surface energy integrand. If, however, the flat curvature flow could be shown to depend continuously on the surface energy, this would give another convergence proof for crystalline approximation as a numerical scheme. As in [16], such an argument would provide no convergence rate.

8. Discussion

We have discussed the crystalline approximation as a numerical scheme for computing the motion of smooth curves by weighted mean curvature. For graphs in the plane we have shown it to be a nonlinear Galerkin scheme, and demonstrated convergence with an error of order $m^{1/2}$ in the H^1 norm. For convex closed curves we have shown it to be a finite difference scheme, and demonstrated convergence with an error of order $(\Delta\theta)^2$ in the Hausdorff distance. These results show, in particular, that the definition of crystalline motion by weighted mean curvature is "correct."

Our methods are, unfortunately, rather special. For example, our analysis of closed curves in the plane requires convexity because it uses the curvature-angle representation $\omega(\theta, t)$. The analysis of nonconvex closed curves seems to require a different method.

Our problems are rather special as well. In defining crystalline motion, we implicitly assumed that no new faces are ever created. Our convergence results justify this assumption. However such an assumption seems to be false for closely related problems such as $V = \omega + g$. To achieve an efficient numerical scheme one needs some guidance concerning when and how to create new faces. We refer to [29, 30] for discussion of a successful implementation, and [28] for some analysis. We mention in passing that crystalline approximation has begun to be used for other problems of phase interface motion, including Saffman-Taylor fingering [4] and surface diffusion [11].

Another issue arises for the motion of two-dimensional surfaces in $\mathbb{R}^3$. If the crystalline solution ceases to exist in the classical sense then our whole approximation scheme breaks down. This situation requires an efficient method of representing varifold solutions. It is not at all clear how this should be done in general; see [34] for a report on recent progress.

Returning to the 1D models considered here, we note an unpleasant feature of crystalline approximation: it is intrinsically piecewise linear. The smooth solution is always approximated by a polygonal one, a representation much more coarse than one could get using front-tracking with splines. We wonder whether there might be a related method which approximates the surface by smoother curves. Perhaps when expressed in curvature-angle coordinates it would become a higher-order difference scheme for (6).

Acknowledgements. PMG did the work reported here and in [18, 19] as a graduate student at the Courant Institute. RVK gratefully acknowledges partial support from NSF and ARO.

References

[1] F. Almgren and J.E. Taylor, *Flat flow is motion by crystalline curvature for curves with crystalline energies*, J. Diff. Geom. 42 (1995) 1–22.

[2] F. Almgren, J.E. Taylor, and L. Wang, *Curvature driven flows: a variational approach*, SIAM J. Cont. Opt. 31 (1993) 387–437.

[3] R. Almgren, *Variational Algorithms and pattern formation in dendritic solidification*, J. Comp. Phys. 106 (1993) 337–354.

[4] R. Almgren, *Crystalline Saffman-Taylor fingers*, SIAM J. Appl. Math. 55 (1995) 1511–1535 .

[5] S. Angenent, D. Chopp, and T. Ilmanen, *A computed example of nonuniqueness of mean curvature flow in* $\mathbb{R}^3$, Comm. Part. Diff. Eqns. 20 (1995) 1937–1958.

[6] S. Angenent and M. Gurtin, *Multiphase thermomechanics with interfacial structure 2. Evolution of an isothermal interface*, Arch. Rat. Mech. Anal. 108 (1989) 323–391.

[7] S. Angenent, T. Ilmanen, and J.J.L. Velazquez, *Nonuniqueness in geometric heat flows*, in preparation.

[8] G. Barles and C. Georgelin, *A simple proof of convergence for an approximation scheme for computing motion by mean curvature*, SIAM J. Numer. Anal. 32 (1995) 484–500.

[9] G. Barles, H.M. Soner, and P.E. Souganides, *Front propagation and phase field theory*, SIAM J. Cont. Opt. 31 (1993) 439–469.

[10] E. Ben-Jacob, R. Kupferman, and O. Schochet, *Numerical study of a morphology diagram in the large undercooling limit using a phase field model*, Phys. Rev. E 50 (1994) 1005ff.

[11] W.C. Carter, A.R. Roosen, J.W. Cahn and J.E. Taylor, *Shape evolution by surface diffusion and surface attachment limited kinetics on completely faceted surfaces*, Acta Metall. et Mater. 43 (1995) 4309–4324.

[12] Y.-G. Chen, Y. Giga, and S. Goto, *Uniqueness and existence of viscosity solutions of generalized mean curvature flow equations*, J. Diff. Geom. 33 (1991) 749–786.

[13] L.C. Evans, *Convergence of an algorithm for mean curvature flow*, Indiana Univ. Math. J. 42 (1993) 533–557.

[14] L.C. Evans, H.M. Soner, and P.E. Souganides, *Phase transitions and generalized motion by mean curvature*, Comm. Pure Appl. Math. 45 (1992) 1097–1123.

[15] L.C. Evans and J. Spruck, *Motion of level sets by mean curvature*, J. Diff. Geom. 33 (1991) 635–681.

[16] T. Fukui and Y. Giga, *Motion of a graph by nonsmooth weighted curvature*, in Proc. First World Congress of Nonlinear Analysts, V. Lakshmikantham, ed., W. de Gruyter, 1995, Vol. I, 47–56.

[17] M. Gage and R.S. Hamilton, *The heat equation shrinking convex plane curves*, J. Diff. Geom. 23 (1986) 69–95.

[18] P.M. Girão, *Convergence of a crystalline algorithm for the motion of a simple closed convex curve by weighted curvature*, SIAM J. Numer. Anal. 32 (1995) 886–899.

[19] P.M. Girão and R.V. Kohn, *Convergence of a crystalline algorithm for the heat equation in one dimension and for the motion of a graph by weighted curvature*, Numer. Math. 67 (1994) 41–70.

[20] M. Gurtin, Thermomechanics of Evolving Phase Boundaries, Oxford Univ. Press (1993).

[21] T.Y. Hou, J.S. Lowengrub, and M.J. Shelley, *Removing the stiffness from interfacial flows with surface tension*, J. Comp. Phys. 114 (1994) 312–338

[22] R. Kobayashi, *A numerical approach to three-dimensional dendritic solidification*, Experimental Math. 3 (1994) 59–81.

[23] B. Merriman, J. Bence, and S. Osher, *Motion of multiple junctions: a level set approach*, J. Comp. Phys. 112 (1994) 334–363.

[24] R.H. Nochetto, M. Paolini, and C. Verdi, *A dynamic mesh algorithm for curvature dependent evolving interfaces*, J. Comp. Phys. 123 (1996) 296–310.

[25] R.H. Nochetto and C. Verdi, *Convergence past singularities for a fully discrete approximation of curvature driven interfaces*, submitted to SIAM J. Numer. Anal.

[26] S. Osher and J. Sethian, *Fronts propagating with curvature dependent speed*, J. Comp. Phys. 79 (1988) 12–49.

[27] S. Roberts, *A line element algorithm for curve flow problems in the plane*, J. Austral. Math. Soc. Ser. B 35 (1993) 244–261.

[28] A.R. Roosen, *Crystalline curvature and flat flow in a diffusion field*, preprint.

[29] A.R. Roosen and J. Taylor, *Simulation of crystal growth with faceted interfaces*, in Mater. Res. Soc. Symposium Proceedings Vol. 237 (1992) 25–36.

[30] A.R. Roosen and J. Taylor, *Modeling crystal growth in a diffusion field using fully faceted interfaces*, J. Comp. Phys. 114 (1994) 113–128.

[31] J. Sethian, *A review of recent numerical algorithms for hypersurfaces moving with curvature-dependent speed*, J. Diff. Geom. 31 (1990) 131–161

[32] H.M. Soner, *Ginzburg-Landau equation and motion by mean curvature, I. Convergence; II. Arbitrary initial data*, J. Geom. Anal., in press.

[33] J.E. Taylor, *Constructions and conjectures in crystalline nondifferential geometry*, in Differential Geometry, Pitman Monographs Surveys Pure Appl. Math. 52, Longman Scientific (1991) 321–336.

[34] J.E. Taylor, *Motion by crystalline curvature*, in Computing Optimal Geometries, J.E. Taylor, ed., Selected Lectures in Mathematics, Amer. Math. Soc., Providence (1991) 63ff.

[35] J.E. Taylor, *Mean curvature and weighted mean curvature*, Acta Metall. et Mater. 40 (1992) 1475–1485.

[36] J.E. Taylor, *Motion of curves by crystalline curvature, including triple junctions and boundary points*, in Differential Geometry, R. Greene and S.T. Yau, eds., Proc. Symp. Pure Math. Vol. 54, Amer. Math. Soc. (1993) 417–438.

[37] J.E. Taylor, J. Cahn, and C. Handwerker, *Geometric models of crystal growth*, Acta Metall. et Mater. 40 (1992) 1443–1474.

[38] A.R. Umantsev, V.V. Vinogradov, and V.T. Borisov, *Modeling the evolution of a dendritic structure*, Sov. Phys. Crystallogr. 31 no. 5 (1986) 596–599.

[39] N.J. Walkington, *Algorithms for computing motion by mean curvature*, SIAM J. Numer. Anal., in press.

[40] A.A. Wheeler, W.J. Boettinger, and G.B. McFadden, *Phase field model for isothermal phase transitions in binary alloys*, Phys. Rev. A 45 (1992), 7424–7439.

[41] A.A. Wheeler, B.T. Murray, and R.J. Schaefer, *Computation of dendrites using a phase field model*, Physica D 66 (1993) 243–262.

Progress in Nonlinear Differential Equations
and Their Applications, Vol. 25

Uses of Elliptic Approximations in Computer Vision*

JAYANT SHAH

Mathematics Department, Northeastern University, Boston, Mass. 02115, USA

1. Introduction

One of the problems in Computer Vision is recovery of object shapes from noisy images. Associated with this problem is the question of what is a shape and how is it to be represented. Since answers to these questions have to be ultimately tailored to the uses one has in mind, one has to bring into consideration potential applications and with it, the question of practical algorithms for implementation of the theory. Here we are concerned with mainly two-dimensional shapes. Mathematically, an object is simply an open subset in the image domain, characterized in some way. In the real world, what is an object and what is just noise or clutter depends of course on what one is looking for. For example, Figure 1a shows a noiseless synthetic image. It may be reasonable to assume that the objects in the figure are the four squares in the four corners and the two ellipses in the middle. Figure 1b shows a noisy version obtained from the image in Figure 1a by adding Gaussian noise. The signal-to-noise ratio (i.e. the ratio between the standard deviation of the image with noise removed and the standard deviation of the noise) is 1 : 4. The problem is to recover the original objects.

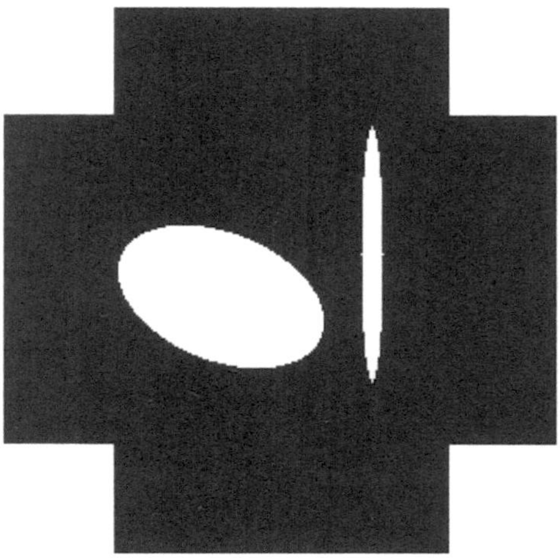

Figure 1a: Noise-free Image

Figure 1b: Noisy Version

*This research was partially supported under ARO Grant, No. DAAL03-91-G-0041.

The general approach is the construction of functionals which incorporate constraints imposed by various objectives such as noise suppression, boundary detection, shape description and object matching. The motivation for building such integrated models comes from a long history of attempts to isolate, formulate and solve simpler problems in Computer Vision and the realization that no matter how sophisticated a technique is, there are inherent ambiguities which cannot be resolved in isolation. It is essential to incorporate contextual information in the model. Moreover, these methods are ideal for formulating models based on raw intensity images. By incorporating the raw image in the model, one can ensure that the input to higher level tasks such as object recognition reflects realistic ambiguities and noise, and thus ensure robustness of the solution. At the same time, integration provides further constraints which may help resolve ambiguities in the image. The disadvantage of such an approach is that as more and more realistic representations are incorporated in these functionals, they become more and more difficult to analyze and implement. In the absence of fast stochastic algorithms, an alternative approach is to construct approximations of these complex functionals, even deliberately weakening the coupling among their components and attempt to find stationary solutions by gradient descent. The purpose of this paper is to describe some numerical experiments involving such approximations.

2. Segmentation Problem

2.1. Segmentation Functional. Traditionally, the first step in image understanding is the detection of edges. The usual method is to smooth the image to reduce noise and then apply some kind of local edge operator such as the zero-crossings of the laplacian. Geman and Geman [GeGe] proposed a Bayesian formulation for simultaneous smoothing and boundary detection. At about the same time, Blake and Zisserman [BZ_1, BZ_2] proposed an analogous discrete model. An analytic version of these models is formulated in [MS_1] as follows:

$$E(u,B) = \frac{1}{\sigma^2}\int_R (u-g)^2 dx\,dy + \int_{R\setminus B} \| \nabla u \|^2 \, dx\,dy + \nu|B| \tag{1}$$

where

R is the image domain, (an open subset in $\mathbf{R}^2$);
g is the feature intensity; $g: R \to \mathbf{R}$;
B is the union of segment boundaries; thus B is the segmenting curve;
$|B|$ is the length of B;
σ and ν are the weights. (σ may be thought of as the smoothing radius in $R \setminus B$.)

The task is to find u and B which minimize $E(u,B)$. Thus the segmentation problem is viewed as the problem of finding a piecewise smooth approximation of g with minimal amount of discontinuity. The weights σ and ν control the relative degrees to which smoothness of u and its discontinuity locus are emphasized. Smoothing is explicitly prevented from extending over B. An object in this formulation is characterized by having relatively uniform feature intensity. Many important theoretical results concerning existence and regularity have been obtained for the special case where R is 2-dimensional and the general case of n-dimensional R, (see for example, [Am], [DMS], [DCL], [MS_2], [Ri], [Sh_4], [Wa]). In particular, in [MS_2] it is shown that the minimal length constraint implies that there cannot be any points where more than three objects meet and an object cannot have corners except at points where three objects meet. At such corners, the angle is always 120°. Thus the formulation implies a very special kind of shape representation and it may be necessary to alter or augment it in order to represent shapes more realistically.

2.2. Dirichlet Version. A simple way to preserve the actual shape singularities is to impose Dirichlet boundary conditions on u in functional (1), namely, $u^{\pm} = g^{\pm}$ along B where the superscript $\pm$ denotes the values on the two sides of B. Results (mostly for the case of 1-dimensional R) concerning the existence and approximations are proved in [Sh_6]. It is shown that in dimension 1 and 2, singularities of u are then a subset of those of g; thus distortion of corners and junctions is prevented. The limitation of this modified model is that if the actual boundaries are noisy (i.e. locally the graph of a noisy function), they will not be smoothed. Moreover, if the boundaries in the image are blurred as is likely in practical situations so that g is continuous, then B would be empty. Hence the formulation must always be used with an explicit representation for blurring. Such a representation based on Ambrosio-Tortorelli approximation (see functional (6) described below) is formulated and analyzed in [Sh_6]. Note that blurring also provides smoothing of noisy boundaries. Recovery of unblurred boundaries from their blurred version is treated as a separate problem.

2.3. Extensions. An obvious extension for enriching shape representation in functional (1) is to incorporate information about the curvature of the boundary. One such extension is proposed in [NM] where $|B|$ is replaced by

$$\int_B (\nu_0 + \nu_1 \kappa^2) ds \tag{2}$$

where κ denotes the curvature. Very few theoretical results concerning the new functional have been obtained. Results for simple closed curves based on elastica for functional (2) under the assumption that $\nu_0 = 0$ and that the length of the curve is fixed have been obtained in [M, W]. In [BDP], lower semi-continuity of functional (2) is analyzed and certain pathologies that arise are exhibited. For instance, even for a simple heart shape, minimizing sequences produce regions of zero width in the limit. Practical difficulties in implementing such a functional are considerable and as yet, no one has implemented it. Allowing B to be only piecewise continuous or introduction of more shape information such as illusory contours or 3D interpretations makes the implementation even more difficult.

2.4. Elliptic Approximation. What makes the task of minimization of functional (1) very difficult is the presence of the one-dimensional segmenting curve. Ambrosio and Tortorelli [AT_1, AT_2] proposed an elliptic approximation of functional (1) so that gradient descent may be applied. Their method is as follows. Let

$$\Lambda_\rho(v) = \frac{1}{2}\int_R \left\{\rho \parallel \nabla v \parallel^2 + \frac{v^2}{\rho}\right\} dx\, dy \tag{3}$$

For a fixed curve B, if v_ρ minimizes $\Lambda_\rho(v)$ with the boundary condition $v = 1$ on B, then as $\rho \to 0$, v_ρ obviously tends to zero everywhere except on B where it equals one. The key observation is that $\Lambda_\rho(v_\rho) \to |B|$ as $\rho \to 0$. Values of v_ρ range between 0 and 1 and v_ρ may be viewed as blurring of B with ρ as the nominal blurring radius. The method of Ambrosio and Tortorelli is to replace the term $|B|$ in functional (1) by $\Lambda_\rho(v)$. B is now spread out over all of R and the integral in functional (1) must be modified as well since we no longer have the set $R \setminus B$. A simple choice is to

$$\text{replace} \quad \int_{R\setminus B} \parallel \nabla u \parallel^2 dx\, dy \qquad by \qquad \int (1-v)^2 \parallel \nabla u \parallel^2 dx\, dy \tag{4}$$

The final form of the approximate functional is

$$E_\rho(u,v) = \int_R \left\{\frac{1}{\sigma^2}(u-g)^2 + (1-v)^2 \parallel \nabla u \parallel^2 + \frac{\nu}{2}\left(\rho \parallel \nabla v \parallel^2 + \frac{v^2}{\rho}\right)\right\} dx\, dy \tag{5}$$

Ambrosio and Tortorelli prove that $E_\rho(u,v)$ converges to $E(u,B)$ in the sense of Γ-convergence. The corresponding approximate functional when Dirichlet boundary conditions are imposed is as follows [Sh$_6$]:

$$\begin{aligned} E_{M,\rho}(u,v) = &\int_R \left[\frac{1}{\sigma^2}(u-g)^2 + (1-v)^2 \parallel \nabla u \parallel^2\right] dx\, dy \\ &+ \int_R \left[\left\{\frac{\nu}{2} + M(u-g)^2\right\}\left\{\left(\rho \parallel \nabla u \parallel^2 + \frac{v^2}{\rho}\right)\right\}\right] dx\, dy \end{aligned} \tag{6}$$

where M is of order $O|\log \rho|$. When $n = 1$, the Γ-convergence of $E_{M,\rho}(u,v)$ to $E(u,B)$ satisfying the boundary conditions $u^\pm = g^\pm$ along B is proved in [Sh$_6$] under certain regularity conditions on g.

It should be noted that an approximation theory for functional (1) when terms containing curvature and singularities of B are inroduced has yet not been found.

Applying gradient descent to (5), we get the following coupled diffusion equations:

$$\begin{aligned} \text{Smoothing}: \frac{\partial u}{\partial t} &= \nabla \cdot (1-v)^2 \nabla u - \frac{1}{\sigma^2}(u-g); \quad \frac{\partial u}{\partial n}|_{\partial R} = 0 \\ \text{Boundary detection}: \frac{\partial v}{\partial t} &= \nabla \cdot \nabla v - \frac{v}{\rho^2} + \frac{2}{\nu\rho}(1-v) \parallel \nabla u \parallel^2; \quad \frac{\partial v}{\partial n}|_{\partial R} = 0 \end{aligned} \tag{7}$$

where ∂R denotes the boundary of R and n denotes the direction normal to ∂R.

The results of Ambrosio-Tortorelli diffusion (7) when applied to the example shown in Figure 1b with $\sigma = \rho = 8$ pixels are shown in Figure 2. The image in Figure 1b was represented on a 256×256 square lattice. The thin ellipse has the maximum width of 9 pixels and the distance between the two ellipses is 18 pixels. Note that long thin objects are very difficult to locate in noisy images because the smoothing radius is governed by the noise characteristics and thus may be much larger than the width of the object. Figure 2a shows the smoothed image u and Figure 2b depicts the edge strength function v. The lighter the area in the figure, the higher the value of v.

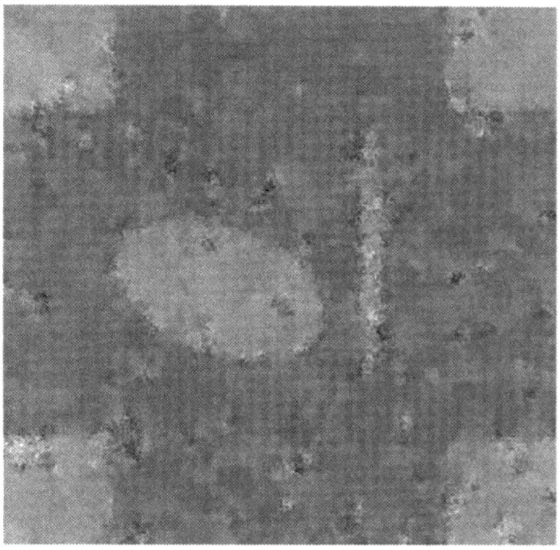

Figure 2a: Smoothed Image u

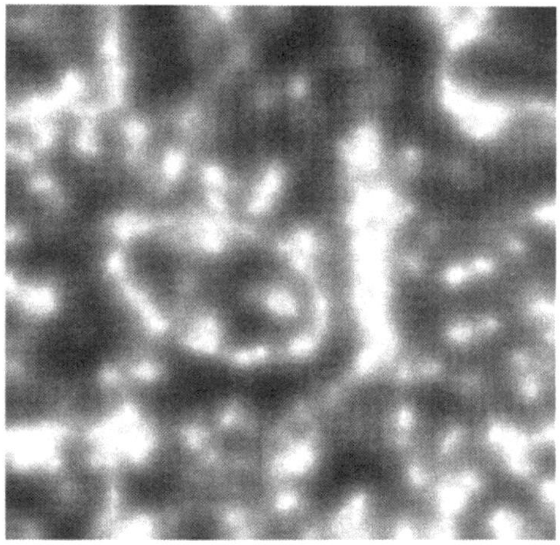

Figure 2b: Edge-strength Function v

3. Shape Recovery

The use of approximate segmentation functionals discussed above for constructing practical algorithms leads to a new difficulty, namely, the difficulty of recovering the actual boundaries from the edge-strength function v. It is apparent from Figure 2b that thresholding of v will not produce satisfactory representation of the boundaries. The trouble is that due to differing levels of contrast along the boundaries and the surrounding noise, values of v along the boundary are not constant. That is, the level curves of v are only approximately parallel to the object boundaries; in fact, they might even be perpendicular to the object boundary in places[1]. Consequently, global thresholding of v produces representation of the boundaries by narrow strips of varying width which may not completely enclose the object if the contrast becomes sufficiently weak. Schemes for adaptive local thresholding of v are discussed in [Sh3]. The point is that the local threshold for v should depend on the average value v in the neighborhood. What this means is that we should

[1]However, it is possible to extract the shape "skeleton" from v which is perhaps more relevant from the point of view of object recognition [TSH]

look at the laplacian of a smoothed version of v. This approach to local thresholding of v does produce improvement, but such further processing was found to produce further displacement of the boundary from its actual location, indicating the need for a better method for recovering the actual boundary from v.

Shape Recovery by Curve Evolution. The basic idea is borrowed from the work of Kass, Witkin and Terzopoulos [KWT] on SNAKES. In their framework, one introduces a simple closed curve in the image and lets it deform under forces which are determined by distance from the object boundary and by a smoothness constaint. In the framework presented here, the analogous idea [Sh_7] is to let an initial curve evolve so as to (locally) maximize the edge-strength function v along the curve in some sense. Let Γ be a simple closed curve in R. In order to move Γ to where the image intensity gradient and hence v are high, we look for the stationary points of the functional

$$L = \int_{\Gamma} (1 - v)^{\alpha} d\gamma \tag{8}$$

where γ denotes the arc-length along Γ. The evolution equation for Γ is derived by applying gradient descent to L. Let $C(p,t) : I \times [0,\infty) \to R$ be the evolving family of curves where I is the unit interval and t denotes time. We require that $C(0,t) = C(1,t)$ for all values of t and that the image of $C(p,0)$ in R coincides with Γ. Then the evolution of Γ is governed by the equation

$$\frac{\partial C}{\partial t} = [\alpha \nabla v \cdot N - (1 - v)\kappa] N \tag{9}$$

where N is the outward normal and κ is the curvature which is defined such that it is positive when Γ is a circle. Thus the points on the evolving curve move in the direction of the normal with velocity which has two components: the component depending on curvature which imposes a smoothness constraint and advection induced by v. The curve is pulled towards the object boundary by the force field ∇v induced by v. The exponent α in the expression for L serves as a weighting factor. The higher the value of α, the weaker the smoothing constraint.

The implementation of curve evolution is no longer a straightforward matter of using finite differences. Moving the points on an evolving curve directly by discretizing the curve leads to many difficulties. For example, the chosen points might bunch up causing numerical instabilities. The curve may also undergo topological changes. An alternate method is the one proposed by Osher and Sethian [OS]. In their approach, the initial curve is embedded in a surface as a level curve and then evolution is applied to the surface so that all of its level curves evolve simultaneously. Assume that Γ is embedded in a surface $f_0 : R \to \mathbf{R}$ as a level curve. Let $f(t,x,y)$ denote the evolving surface such that $f(0,x,y) = f_0(x,y)$. Then, in order to let all the level curves of f_0 evolve simultaneously, we consider the functional

$$G(f) = \int_{-\infty}^{\infty} \int_{\Gamma_c} (1 - v)^{\alpha} d\gamma_c dc \tag{10}$$

where $\Gamma_c = \{(x,y)|f(t,x,y) = c\}$. But by the coarea formula[2],

$$G(f) = \int\int_R (1-v)^\alpha \parallel \nabla f \parallel dx\, dy\ . \tag{11}$$

By calculating the first variation of the last functional, we get the gradient descent equation as

$$\begin{aligned} \frac{\partial f}{\partial t} &= -\alpha \nabla v \cdot \nabla f + (1-v) \parallel \nabla f \parallel \nabla \cdot \left(\frac{\nabla f}{\parallel \nabla f \parallel} \right) \\ &= -\alpha \nabla v \cdot \nabla f + (1-v) \frac{f_y^2 f_{xx} - 2 f_x f_y f_{xy} + f_x^2 f_{yy}}{\parallel \nabla f \parallel^2} \\ \frac{\partial f}{\partial n}|_{\partial R} &= 0;\ f = f_0 \text{ at } t = 0. \end{aligned} \tag{12}$$

An important question now is: How to specify the initial curve Γ? An automatic specification of the initial curve is a difficult problem because its solution implies that the object boundaries have already been found, at least approximately. The strategy used in [Sh_7] for specifying the initial curve is based on the following considerations. If we set $v = 1$ and $f_0 = g$ in equation (12), then it reduces to the diffusion equation proposed in [ALM] for smoothing images. In [ALM], diffusion is allowed to occur only in the direction of the level curves, the expectation being that in this way, the object boundaries will be smoothed without being blurred. The difficulty with this approach is of course that the object boundaries in general are not level curves in an image. Traditional use of the zero-crossings of the laplacian of smoothed images to detect edges may be thought of as an attempt to remedy this situation by representing object boundaries by the level curves of the smoothed laplacian instead of the level curves of the image itself. Hence, the strategy in [Sh_7] is to use the zero-crossing of the laplacian as the initial approximation Γ for the object boundaries. A fundamental difficulty with the use of level curves as object boundaries is that level curves cannot represent triple points which are very important features in image analysis. This problem is not considered here. Another difficulty is that the theory of curve evolution is based on the assumption that Γ is a simple closed curve, an assumption which is violated at the saddle points of f. If the laplacian is too noisy, the evolution is dominated by what happens at its saddle points and becomes unpredictable. This is because at a saddle point, the first term on the right hand side of the evolution equation (12) vanishes and both the numerator and the denominator vanish in the second term. Hence, the behaviour of the second term becomes very sensitive to noise near a saddle point. In order to demonstrate the feasibility of this approach, an ad hoc solution to this problem has been adopted in [Sh_7]. The Ambrosio-Tortorelli solution u is further smoothed by non-uniform smoothing until the zero-crossings of the laplacian of the smoothed version have relatively few self-intersections (see [Sh_7] for more details). Then we set f_0 equal to the laplacian of the smoothed u and let it evolve according

[2] The functional $G(f)$ has been extended to a segmentation functional analogous to functional (1); see [Sh_8].

to equation (12). The zero-crossings of the evolving f are taken as the successive refinements of the object boundaries.

Figure 3: Evolution of zero-crossings

For numerical implementation of equation (12), the usual finite-difference schemes (say, central differences) are exactly the wrong thing to apply and must never be used. The main point of Osher and Sethian is that since we expect the evolving surface to develop discontinuities or "shocks" where the object boundaries are, the directions of the finite differences must be chosen adaptively. Once this principle is incorporated in the numerical scheme, nonlinear diffusion (12) behaves very robustly.

Figures 3 and 4 show the results of evolution when applied to the example shown in Figure 1b. Figure 3 shows the zero-crossings of the evolving laplacian at $t = 0, 40, 80, 160, 320, 640, 1280$. Thus, the zero-crossings at $t = 0$ are the zero-crossings of the laplacian of a smoothed version of u. The example illustrates that the boundaries found by this method are at least metastable, that is, they persist for a long time. Figure 4 shows the superposition of the zero-crossings smoothed by evolution ($t = 1280$) on the noisefree image (Figure 4a) as well as on the smoothed image u (Figure 4b). Smoothed image u was used for superposition rather than g because the evolution is governed by v which, in turn, is determined by the boundaries in u. Note the accuracy of placement of the smoothed zero-crossings on the noiseless image, including the thin ellipse. The corners are also fairly well represented. The boundary deviates in places from the boundary in the noiseless image when it follows some accidental feature introduced by the noise. (It is possible to discern these features by the inspection of Figure 2a.) It is interesting that although the four corner squares are identical in the noiseless image, the final boundaries are different in all four cases because the accidental noise features are different. The worst deviation occurs in the case of the lower right corner square. The very thin ends of the thin ellipse are lost because the boundary representation by v is too coarse ($\rho = 8$ pixels). Some portions towards the ends of the thin ellipse are lost because they are obscured by noise as one can see in Figure 2a.

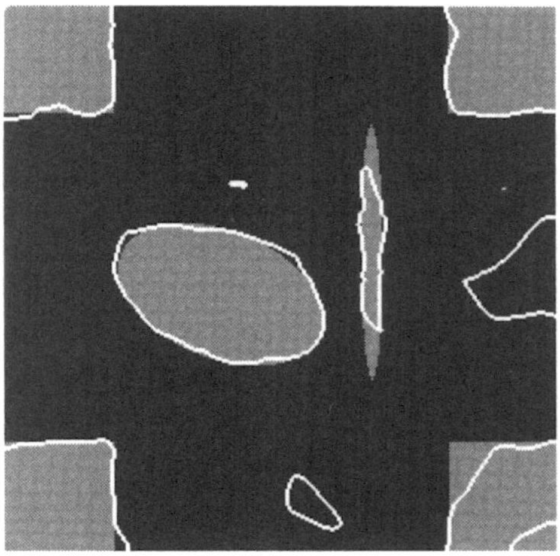

Figure 4a

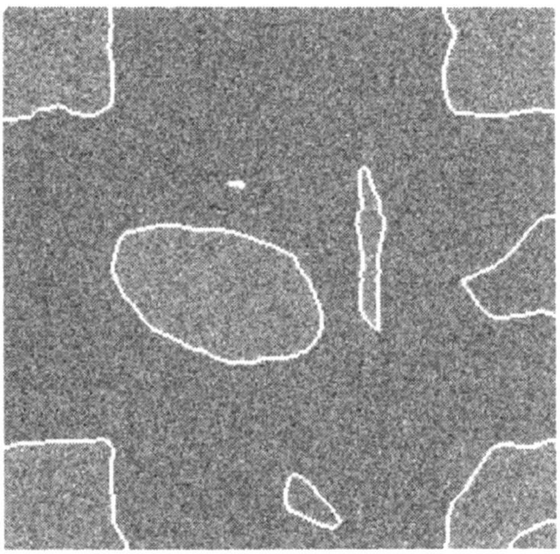

Figure 4b

4. Diffusion Systems

Use of curve evolution for shape recovery as described above exemplifies an approach for getting around the difficulties in finding the minimizing solutions for functionals in Computer Vision. The basic ingredient was to incorporate the edge-strength function v in the algorithms designed to solve the next level of tasks in Computer Vision. If these tasks are also governed by evolution equations, then the idea is to let this evolution be controlled by v. In fully integrated models, one would try to construct a coupled system of diffusion equations in which each equation controls one particular process or a task like smoothing or edge detection or matching and outputs of these equations provide controls for each other's evolution. The diffusion system (7) of Ambrosio and Tortorelli may be viewed as a system of this type. Other examples are nonlinear smoothing [Sh_3, PPGO], stereo [Sh_5] and optical flow [PGO]. The difficulty in designing nonlinear systems of diffusion equations for problems in Computer Vision is that such systems are one of the most difficult mathematical objects to analyze and no general mathematical principles are available to guide the design. The examples cited above are based on purely heuristic arguments. For illustration, the case of stereo is described below in some detail.

A diffusion system for the stereo problem. In stereo vision, the basic problem is the correspondence problem, that is, the problem of matching the two views seen by the two eyes (or cameras) and thus computing the disparity between the two images. Let R_l and R_r be the left and right image planes respectively, assumed to be coplanar. Let $I_l : R_l \to \mathbf{R}$ and $I_r : R_r \to \mathbf{R}$ be the respective image intensities. A major complication in matching the two images is the presence of discontinuities. There are sudden jumps in the disparity at the boundaries between objects which are at differing distances from the eyes. Another consequence of this is the phenomonon of half-occlusions which are the areas in the scene seen by one eye and not the other. Thus it is necessary to prevent matching of the two images in areas of half-occlusions. In the one-dimensional case, these considerations lead to the following functional for the map $f_l : R_l \to R_r$ matching the left image with the right image:

$$\begin{aligned} E_l(f_l) = &\int_{R_l \setminus (B_l \cup O_l)} \left\{ (f_l')^2 + \left(\frac{1}{f_l'}\right)^2 + \frac{4}{\sigma^2}[I_l - I_r \circ f_l]^2 \right\} dx_l \\ &+ 4 \sum_{x_l \in B_l} (\nu + \gamma |f_l^+(x_l) - f_l^-(x_l|) \\ &+ 4\nu(\text{ number of components of } O_l) + 4\gamma \text{ length } (O_l) \end{aligned} \tag{13}$$

where $B_l = \{x_l : f_l'(x_l) = \infty\}$ and occlusion $O_l = \{x_l : f_l'(x_l) = 0\}$

A similar functional may be obtained for the map $f_r : R_r \to R_l$ matching the right image with the left image. Maps f_l and f_r are inverse of each other. In particular, there is a correspondence between the components of B_l and the components of O_r. Similarly, there is a correspondence between the components of B_r and the components of O_l. Complexity of functional (13) forces us to look for approximate solutions. For this purpose, orient coordinates x_l in R_l and x_r in R_r from left to right and define the left disparity d_l by the equation

$$f_l = x_l + d_l \tag{14}$$

Then,

$$(f_l')^2 + \left(\frac{1}{f_l'}\right)^2 \approx 2 + 4(d_l')^2 \text{ if } |d_l'| << 1 \tag{15}$$

Define a functional $\hat{E}_l(d_l)$ over R_l as follows:

$$\begin{aligned}\hat{E}_l(d_l) = &\int_{R_l \setminus (B_l \cup O_l)} \left\{ (d_l')^2 + \frac{1}{\sigma^2}[I_l - I_r \circ f_l]^2 \right\} dx_l \\ &+ \sum_{x_l \in B_l} (\nu + \gamma |f_l^+(x_l) - f_l^-(x_l)|) \\ &+ \nu(\text{ number of components of } O_l) + \gamma \text{ length } (O_l)\end{aligned} \tag{16}$$

Similarly, define d_r and $\hat{E}_r(d_r)$ after orienting x_l and x_r from right to left. Then

$$E_l(f_l) = C + 4\hat{E}_l(d_l) \approx C + 4\hat{E}_r(d_r) = E_r(f_r) \tag{17}$$

where C is a constant. Hence we consider the functionals $\hat{E}_l(d_l)$ and $\hat{E}_r(d_r)$ in order to find approximate solutions. Except for the presence of O_l and O_r, functionals $\hat{E}_l$ and $\hat{E}_r$ are very similar in form to the functionals used in segmentation problems. Since $B_l \cap O_l$ and $B_r \cap O_r$ are empty, a strategy to find approximate solutions would be as follows: Apply gradient descent alternately to $\hat{E}_l$ and $\hat{E}_r$. That is, alternately minimize $\hat{E}_l$ with respect to d_l and B_l keeping O_l fixed and minimize $\hat{E}_r$ with respect to d_r and B_r keeping O_r fixed. Determination of the discontinuity locus B_l when $\hat{E}_l$ is minimized also determines the occlusion O_r in R_r since $O_r = R_r \setminus$(range of f_l). Similarly the discontinuity locus B_r determines the occlusion O_l in R_l.

We make three further simplifications before implementing this strategy. First, we follow the example of segmentation functionals and simplify the penalty term for the discontinuity locus by setting $\gamma = 0$. Next, we drop the constraint that f_l and f_r have the same graph. This is based on the heuristic that minimization of $\hat{E}_l$ and $\hat{E}_r$ without this constraint will still produce comparable optimal values for disparities. Finally we simplify the constraint that f_l and f_r must extend continuously across the occlusion sets O_l and O_r respectively such that they are constant over the occlusion set. We relax this requirement by merely extending d_l and d_r across the occlusion set and including the cost of $|d_l'|^2$ and $|d_r'|^2$ over the occlusion set in their respective functionals. The 2-dimensional generalization is straight-forward. The final result is a pair of coupled functionals, E_l and E_r

defined over R_l and R_r respectively as follows:

$$
\begin{aligned}
E_l(d_l) &= \int\int_{R_l \setminus B_l} \| \nabla d_l \|^2 + \frac{1}{\sigma^2} \int\int_{R_l \setminus O_l} \{[I_l - I_r \circ f_l]^2\} + \nu |B_l| \\
E_r(d_r) &= \int\int_{R_r \setminus B_r} \| \nabla d_r \|^2 + \frac{1}{\sigma^2} \int\int_{R_r \setminus O_r} \{[I_r - I_l \circ f_r]^2\} + \nu |B_r|
\end{aligned} \tag{18}
$$

where $O_l = R_l \setminus (\text{range of } f_r)$ and $O_r = R_r \setminus (\text{range of } f_l)$

In this form, the functionals are also valid in the 2-dimensional case. For a practical implementation by gradient descent, we now mimick the method of Ambrosio and Tortorelli and approximate the discontinuity loci B_l and B_r by continuous functions w_l and w_r. To calculate w_l corresponding to given d_l, apply gradient descent to the functional

$$
E_D(w_l) = \int\int \left\{ \rho \| \nabla w_l \|^2 + \frac{w_l^2}{\rho} + 2(1 - w_l)^2 \| \nabla f_l \|^2 \right\} \tag{19}
$$

To calculate d_l for given w_l, f_r and w_r, apply gradient descent to the functional

$$
E_m(d_l) = \int\int \left\{ (1 - w_l)^2 \| \nabla d_l \|^2 + \frac{1}{\sigma^2} (I_l - I_r \circ f_l)^2 (1 - w_r \circ f_r^{-1})^2 \right\} \tag{20}
$$

Similarly, define $E_d(w_r)$ and $E_m(d_r)$ to calculate w_r and d_r. The result is a coupled system of 4 diffusion equations as follows:

With x_l oriented from left to right, equations for d_l and w_l are

$$
\begin{aligned}
\frac{\partial d_l}{\partial t} &= \nabla \cdot \left((1 - w_l)^2 \nabla d_l \right) \\
&\quad + \frac{1}{\sigma^2} [I_l - I_r \circ f_l] \left(\frac{\partial I_r}{\partial x_r} \circ f_l \right) (1 - w_r \circ f_r^{-1})^2 \\
\frac{\partial w_l}{\partial t} &= \rho \nabla^2 w_l - \frac{w_l}{\rho} + \frac{2}{\nu} (1 - w_l) \| \nabla f_l \|^2
\end{aligned} \tag{21}
$$

With x_r oriented from right to left, equations for d_r and w_r are

$$
\begin{aligned}
\frac{\partial d_r}{\partial t} &= \nabla \cdot \left((1 - w_r)^2 \nabla d_r \right) \\
&\quad + \frac{1}{\sigma^2} [I_r - I_l \circ f_r] \left(\frac{\partial I_l}{\partial x_l} \circ f_r \right) (1 - w_l \circ f_l^{-1})^2 \\
\frac{\partial w_r}{\partial t} &= \rho \nabla^2 w_r - \frac{w_r}{\rho} + \frac{2}{\nu} (1 - w_r) \| \nabla f_r \|^2
\end{aligned} \tag{22}
$$

The boundary conditions are as follows. At the ends of each epipolar line, we require that d_l and d_r must be zero. There is no loss of generality by this assumption. If the disparities in fact are not zero at the end-points, then the situation is represented by making disparity discontinuous at such end-points and having a corresponding occlusion area touching the end-points in the image of the other eye. The remaining boundary conditions are the homogeneous Neumann boundary conditions; that is, the normal derivative of the variables is set equal to 0.

For illustration, we present an example of a random dot stereogram shown in Figure 5. Each "dot" is 4×4 pixel in size. Each image consists of 256×256 pixels. The left image is obtained from a random-dot image by shifting the central 96×96 pixel square to the left by 8 pixels and shifting the 176×176 square surrounding it by 4 pixels to the left. The right image is obtained by shifting the same squares to the right by the same amounts. Thus the reconstructed object is a square "wedding cake". Figure 6 portrays the numerically computed disparity values. The brighter areas indicate the higher disparity values. Figure 7 depicts the values of w_l and w_r. The bright areas indicate the blurred version of the discontinuity locus of d_l and d_r while the dark areas correspond to occlusion. The uniform gray indicates approximately constant disparity.

Figure 5: Left and right images

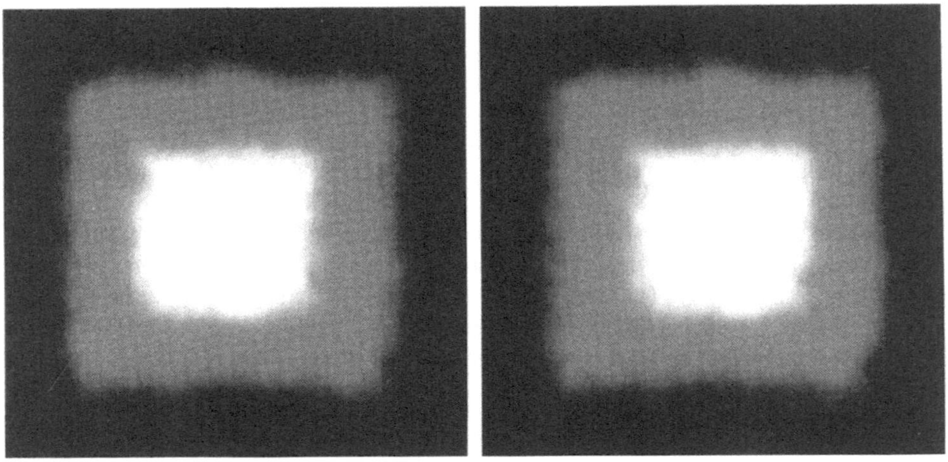

Figure 6: Left and right disparities d_l, d_r

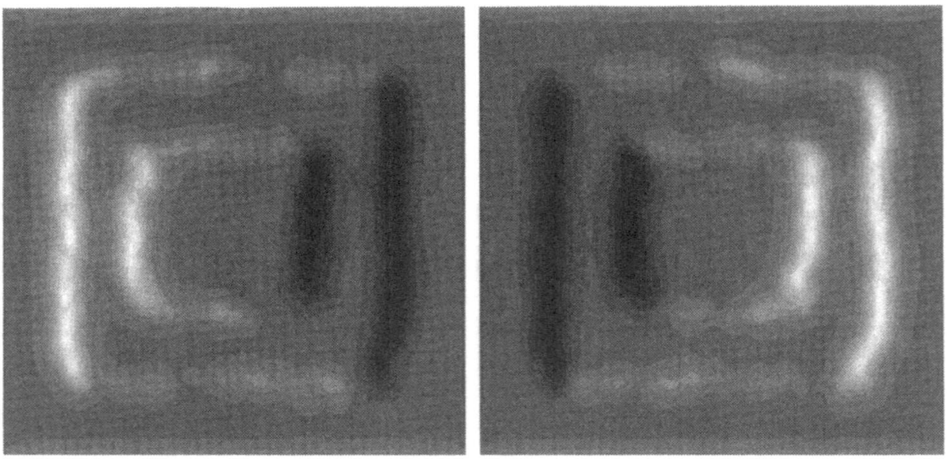

Figure 7: w_l, w_r

Bibliography

[ALM] L. Alvarez, P.L. Lions and J.M. Morel: "Image Selective Smoothing and Edge Detection by Nonlinear Diffusion II", SIAM J. Num. Anal., (June, 1992).

[AT_1] L. Ambrosio and V.M. Tortorelli: "Approximation of Functionals Depending on Jumps by Elliptic Functionals via Γ-convergence", Arch. Rat. Mech. Anal. 111, pp. 291–322, (1990).

[AT_2] L. Ambrosio and V.M. Tortorelli: "On the Approximation of Functionals Depending on Jumps by Quadratic, Elliptic Functionals", Boll. Un. Mat. Ital. (1992).

[BDP] G. Bellettini, G. Dal Maso and M. Paolini: "Semicontinuity and Relaxation Properties of a Curvature Depending Functional in 2d", SISSA Report, 17/92/MA, (Feb. 1992).

[BZ_1] A. Blake and A. Zisserman: "Using weak continuity constraints", Report CSR-186-85, Dept. of Comp. Sci., Edinburgh Univ. (1985).

[BZ_2] A. Blake and A. Zisserman: Visual Reconstruction, M.I.T. Press (1987).

[DMS] G. Dal Maso, J.M. Morel and S. Solimini: "A Variational Method in Image Segmentation: Existence and Approximation Results", Acta Math. 168, pp. 89–151, (1989).

[DCL] E. De Giorgi, M. Carriero and A. Leaci: "Existence Theorem for a Minimum Problem, with Free Discontinuity Set", Arch. Rat. Mech. Anal. 108, pp. 195–218, (1990).

[GeGe] S. Geman and D. Geman: "Stochastic relaxation, Gibbs' distributions, and the Bayesian restoration of images", IEEE Trans., PAMI 6, pp. 721–741, (1984).

[KWT] M. Kass, A. Witkin and D. Terzopoulos: "Snakes: Active Contour Models", First International Conf. on Computer Vision, (1987).

[M] D. Mumford: "Elastica and Computer Vision", Preprint.

[MS_1] D. Mumford and J. Shah: "Boundary detection by minimizing functionals, I", Proc. IEEE Conf. on Computer Vision and Pattern Recognition, San Francisco, (1985).

[MS_2] D. Mumford and J. Shah: "Optimal approximations by piecewise smooth functions and associated variational problems", Comm. on Pure and Appl. Math., v. XLII, n. 5, pp. 577–684 (July, 1989).

[NM] M. Nitzberg and D. Mumford: "The 2.1D Sketch", Third Int'l Conf. on Comp. Vision, (December, 1990).

[OS] S. Osher and J. Sethian: "Fronts Propagating with Curvature Dependent Speed: Algorithms based on the Hamilton-Jacobi Formulation", J. Comp. Physics, 79, (1988).

[PPGO] M. Proesmans, E.J. Pauwels, L.J. Van Gool and A. Oosterlink: "Image Enhancement using Non-Linear Diffusion", Proc. IEEE Conf. on Comp. Vision and Pattern Recognition, (June, 1993).

[PGO] M. Proesmans, L.J. Van Gool and A. Oosterlink: "Determination of Optical Flow and its Discontinuities using Non-linear Diffusion", Preprint, ESAT-MI2, Katholike Univ. Leuven, Leuven, Belgium, (1993).

[Ri] T. Richardson: Ph.D. Thesis, Department of Electrical Engineering and Computer Science, MIT (1990).

[Sh_1] J. Shah: "Parameter Estimation, Multiscale Representation and Algorithms for Energy-Minimizing Segmentations", Tenth International Conference on Pattern Recognition, (June, 1990).

[Sh_2] J. Shah: "Segmentation by Nonlinear Diffusion", Proc. IEEE Conf. on Comp. Vision and Pattern Recognition, (June, 1991).

[Sh_3] J. Shah: "Segmentation by Nonlinear Diffusion, II", Proc. IEEE Conf. on Comp. Vision and Pattern Recognition, (June, 1992).

[Sh_4] J. Shah: "Properties of Energy-Minimizing Segmentations", SIAN J. on Control and Optim. v. 30, no. 1, 99–111, (1992).

[Sh_5] J. Shah: "A Nonlinear Diffusion Model for Discontinuous Disparity and Half-Occlusions in Stereo", Proc. IEEE Conf. on Comp. Vision and Pattern Recognition, (June, 1993).

[Sh_6] J. Shah: "Piecewise Smooth Approximations of Functions", Calculus of Variations and Partial Differential Equations, 2, pp. 315–328, (1994).

[Sh_7] J. Shah: "Shape Recovery from Noisy Images by Curve Evolution", IASTED International Conference on Signal and Image Processing, (Nov. 1995).

[Sh_8] J. Shah: "A Common Framework for Curve Evolution, Segmentation and Anisotropic Diffusion", Proc. IEEE Conf. on Computer Vision and Pattern Recognition, San Francisco (1996).

[TSH] S. Tari, J. Shah and H. Pien: "A Computationally Efficient Shape Analysis via Level Sets", IEEE Workshop on Mathematical Methods in Biomedical Image Analysis (1996).

[Wa] Y. Wang: Ph.D. Thesies, Dept. of Math., Harvard Univ., (1989).

[W] Y. Wen: "L^2 Flow of Curve Straightening in the Plane", Duke Math. J. v. 70, n. 3, pp. 683–398, (June, 1993).

Progress in Nonlinear Differential Equations
and Their Applications, Vol. 25

A Kanizsa Programme

VICENT CASELLES*, BARTOMEU COLL*, JEAN-MICHEL MOREL*

* Departament de Matemàtiques i Informàtica
Univ. Illes Balears, 07071 Palma (Balears), Spain
* Ceremade, Univ. Paris Dauphine, Pl. de Lattre de Tassigny,
75776 Paris cedex 16, France

ABSTRACT. We discuss the physical generation process of images as a combination of occlusions, transparencies and contrast changes. This description fits to the phenomenological description of Gaetano Kanizsa, according to which visual perception tends to remain stable with respect to these basic operations by detecting several kinds of essential singularities: T-junctions and X-junctions, generated respectively by occlusion and transparency. We deduce a mathematical and computational model for image analysis according to which the "atoms", or starting elements of every image analysis process must be "pieces of level lines joining T-or X-junctions". An elementary and rigorous algorithm, parameter-free except for a threshold eliminating quantization effects, is proposed for the computation of the "atoms of perception" thus defined.

1. Introduction

This paper is devoted to the analysis of singularities in a both exotic and familiar material: the digital images. It can be said without much exageration that the scientific and technical discipline called *image processing* has devoted most of its effort to the "edge detection" problem, which is nothing but a discontinuity detection problem. The edges have been and are considered as the basic objects in images, the "atoms" on which all Computer Vision algorithms can be built [Ma]. There are many possible approaches to the edge detection problem and many techniques from functional analysis have been proposed. A review of both the variational and the PDE approaches can be found in [MS]. To make short a long story, let us recall first that a digital image is modelized as a real function $u(x)$, where x represents an arbitrary point of the plane and $u(x)$ denotes the grey-level at x. In practice, an image has discontinuities everywhere, so that some selection process of the "true" discontinuities must be defined. One way to do this selection is to have an *a priori model* of the image, describing which kind of discontinuities are expected (and which kind of regularity). These expectations are translated into an energy functional $E(u, u_0)$ where u_0 is the original digital image, and u an arbitrary element of an admissible class of interpretable images (e.g. with smooth regions and smooth discontinuity lines). Such a model for images is to impose (as proposed by Rudin in 83) to u that it belongs to BV (space of functions with bounded variation), so that, by a classical theorem in geometric measure theory, the discontinuity set is contained in a countable set of rectifiable curves.

Another way to the selection of the "right" discontinuities is to smooth previously the image by some convolution or diffusion process, after which edges are detected as local extrema of the gradient. Then these points must be connected to form curves by a geometric measure criterion, so that, at the end, all methods of edge detection are variational. We refer to [MS] again for an review and a complete mathematical discussion of PDEs and variational methods in image processing.

Our aim here is to present an alternative definition of the basic objects of image processing. We shall argue that only "pieces of level lines of the image joining T- or X-junctions" can be considered as the atoms of visual perception, on which further algorithms can be built. There is no full contradiction in this view with the edge detection ideology. Now, a consequence of our discussion is that edge detection, as described above, is no "low level" process of vision, be it artificial or natural.

Our argumentation is as follows. First, we describe by a very simple physical model what the image generation process is. Then we deduce from invariance requirements with respect to the accidents of this generation process what the atoms of visual perception must be. The main accidents in the generation process are: occlusion, transparency and contrast change. On the other hand, all of our argumentation can be read as a computational-mathematical explanation of Gestaltists theories about several salient facts of our perception [Ar]. In particular, a result of our inquiry is to give a computational model for the pregnant singularities of Kanizsa [Ka1, 2, 3] theories: T-junctions, transparencies (X-junctions). We show that there is a single rigorous way for computing them and show experiments on real images. The algorithms are extremely simple, since they are based on the computation of level lines as the topological boundaries of level sets (which are computed by a simple thresholding !). In particular, the algorihms involve no parameters (except for an obvious threshold on occlusions necessitated by the quantization effects of the digital image). The "atoms" thus found can be used as starting objects for the reconstruction of depth in the image by variational method, as proposed by Nitzberg and Mumford in their pioneering work [NM]. Among the previous works which have considered algorithms for the detection of T-junctions in images, we wish to mention [A], which combines edge detection techniques and mathematical morphology. Earlier attempts to base the detection of T-junctions on a previous edge detection are proposed in [DG], [Li] and [NM]. These methods are all based on a Gaussian-like convolution followed by an analysis of edges and are not invariant with respect to contrast changes, which makes them basically different from what we propose. In fact, the detection of T-junctions based on a previous smoothing of the image tends to destroy the T-junctions [DG, AM], thus necessitating a subsequent following up of the edges to restore them. We shall show that the classical dogma "filtering must be done before sampling" is not adapted in this case. In [AM] is presented a rigorous theory for detecting corners which seems optimal and in some extent completes our discussion. The method presented therein is indeed invariant with respect to constrast changes.

2. How images are generated: occlusion and transparency as basic operations

Before starting with the discussion on the basic objects of image processing, we wish to discuss briefly the achievements of signal analysis, that is, the numerical analysis of sounds (music, voice, etc.). This discussion will be very useful as an example of rigorous and physically well-founded approach to signal processing. In acoustics, the main operation by which sounds are generated is the periodic vibration of some rigid or elastic material. This explains why the notion of frequency (or local frequency) has so much relevance. From the mathematical viewpoint, elementary sounds are oscillatory signals and when several sounds are simultaneously emitted, the resulting signals simply add. Thus, in a first approximation, the world of sounds is a linear space and linear decompositions, invariant with respect to time-shifts, are therefore canonically associated with sound analysis. Whence the importance of linear, translation invariant operators like convolutions in general and of Fourier analysis and its variants in particular: windowed Fourier, Gabor transforms, time-frequency analysis, wavelet decompositions, etc. in everything related to the analysis of voice, music, and wave propagation in general. Indeed, the natural (linear) decomposition of a sound into elementary objects tends to split locally the sound into more elementary sounds which can be better localized in frequency (Fourier) or in time (wavelets), or both (Gabor, wavelet packets) and as different as possible. The Gabor analysis explicitly aimed at an automatic musical notation [Me].

In the same way, Vision theory and Image analysis should have their basic objects and those basic objects should be deduced from the way images have been generated. This has little to do with the wave nature of light propagation. Indeed, frequency, or wavelength, is taken into account in a rather rough way as *color* and the only analogy we can draw between acoustics and image analysis in this direction should point to the way new colours are obtained by combining (adding) waves, with several wavelengths. In fact, the basic model for image generation is essentially geometric and only assumes that light propagates in straight lines, a physically approximately true law. According to this basic model, the *camera oscura* or *pinhole* model, every point of the surface of physical objects reflects solar or artificial light in all directions. Generation of images occurs when this light is focussed by passing through a hole on the side of a black box and finally impressed on the face of the black box opposite to the hole. The black box can be a camera, in which case this face is a photosensitive paper or an array of electronic amplifiers. It also is the model of the eye, in which case the photosensitive surface is the retina. (See Figure 1).

Calling x a generic point on the retina (or film) plane, we call $u(x)$ the result of this impression, which we shall throughout this paper assume to be a real value measuring intensity (or energy) of the light sent onto x. It is sometimes assumed that light sent by the surface of objects is the same in every direction (the so called *Lambertian assumption*). This assumption is, however, false when mirroring effects occur. In any case, since the repartition of the light sources is not uniform, it is expected that different parts of the surface of physically homogeneous objects will reflect different light intensities. In addition, objects may occult to each

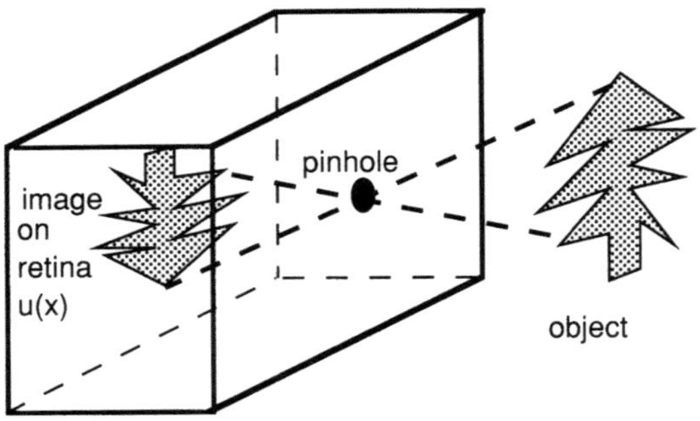

Figure 1: Camera oscura (black box)

other some light sources, so that shadows appear on their surface. The intricacy of natural images is therefore extreme, and we certainly cannot pretend that images faithfully reflect the objects appearing in the world. In fact, light sources, shadows, reflectances, apparent contours are the basic visual objects, more basic for sure than the real objects. Our main concern here is to find how those basic objects can be used to extract physical information on objects from images. In order to do that, we must follow the same method as in acoustics; we must deduce the structure of the basic objects from the way they have been generated. We shall, following the psychologist and gestaltist Gaetano Kanizsa, define two basic operations for image generation: *occlusion and transparency.* In the same way as in acoustics, where the basic operation, the superposition of transient waves is interpreted as an addition of functions in a Hilbert space, we shall interpret occlusion and transparency as basic operations on images considered as functions $u(x)$ defined on the plane. Before beginning with the discussion about geometric operations on images, we wish to make clear the role which, in our opinion, linear methods play in image processing. Most optical devices operate a convolution on the signal and add random perturbations ("noise") to the original image. Thus, most early image processing methods have focussed on linear decomposition methods for image restoration, which is sound since the initial perturbations also are linear in a first approximation. We shall not adress this matter here and we shall focus on image analysis methods where either those perturbations are negligible, or they already have been corrected so that we can directly adress the geometric structure of images.

2.1. Occlusion as a basic operation on images Given an object $\tilde{A}$ in front of the camera, we call A the region of the plane to which it corresponds by the pinhole model, that is, the region it covers on the "retina". We call u_A the image thus generated, which is defined in A. Assuming now that the object $\tilde{A}$ is added in a real scene $\tilde{R}$ of the world whose image was v, we observe a new image which depends upon which part of $\tilde{A}$ is in front of objects of $\tilde{R}$, and which part in back. Assuming that $\tilde{A}$ occludes objects of $\tilde{R}$ and is occluded by no object of $\tilde{R}$, we get a new image $u_{\tilde{R}\cup\tilde{A}}$ defined by

$$\begin{aligned} u_{\tilde{R}\cup\tilde{A}} &= u_A \text{ in } A \text{ and} \\ u_{\tilde{R}\cup\tilde{A}} &= v \text{ in } \mathbb{R}^2 \setminus A. \end{aligned} \tag{1}$$

Of course, we do not take into account in this basic model the fact that objects in $\tilde{R}$ may intercept light falling on $\tilde{A}$, and conversely. In other terms, we have omitted the shadowing effects, which will now be considered.

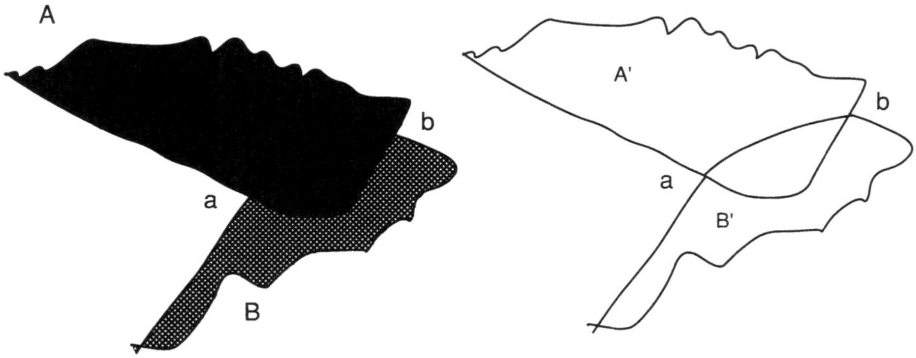

Figure 2: T-junctions and occlusions: the left-hand figure, composed of two regions A and B is interpreted as the superposition of two objects A′ and B′ which are represented on the right by their contours. The completion thus effectuated by a low level vision process consists in extending beyond the T-junctions a and b the boundary of the object B′, which apparently "stops" when it meets the object A′. This interruption is interpreted as an occlusion.

2.2. Transparency (or shadowing) as a second basic operation on images. Let us assume that the light source is a point in euclidian space, and that an object $\tilde{A}$ is interposed between a scene $\tilde{R}$ whose image is v and the light source. We call $\tilde{S}$ the shadow spot of $\tilde{A}$ and S the region it occupies in the image plane. The resulting image u is defined by

$$\begin{aligned} u_{\tilde{R},\tilde{S},g} &= v \text{ in } \mathbb{R}^2 \setminus S \text{ and} \\ u_{\tilde{R},\tilde{S},g} &= g(v) \text{ in } S. \end{aligned} \tag{2}$$

Here, g denotes a constrast change function due to the shadowing, which is assumed to be uniform in $\tilde{S}$. Clearly we must have $g(s) \leq s$, because the brightness decreases inside a shadow, but we do not know in general how g looks. The only assumption for introducing g is that points with equal grey level s before shadowing get a new, but the same, grey level $g(s)$ after shadowing. Of course, this model is not true on the boundary of the shadow, which can be blurry because of diffraction effects or because the light source is not really reducible to a point. Another problem is that g in fact depends upon the kind of physical surface which is shadowed so that it may well be different on each one of the shadowed objects. This is no real restriction, since this only means that the shadow spot S must be divided into as many regions as shadowed objects in the scene, so that we only need to iterate the application of the preceding model accordingly. A variant of the shadowing effect which has been discussed in perception psychology by Gaetano Kanizsa is *transparency*. In the transparency phenomenon, a transparent homogeneous object $\tilde{S}$ (in glass for instance) is interposed between part of the scene and the observer. Since $\tilde{S}$ intercepts part of the light sent by the scene, we still get a relation like (2), so that transparency and shadowing simply are equivalent from the image processing viewpoint. If transparency (or shadowing) occurs uniformly on the whole scene, the relations (2) simply become

$$u_g = g(v),$$

which means that the grey-level scale of the image is altered by a nondecreasing contrast change function g. In natural or artificial vision, this happens all the time because solar light is constantly changing intensity, being obscured by aerosols or clouds while the light captors themselves (no matter artificial or natural) are subject to sensibility changes. There is, however, a difference between the change of contrast sensibility of a receptor and a change in the light sources intensity. Indeed, in this last case, the response of different objects to this change may be different (according to their material) so that we may expect that the resulting constrast change g depends upon the objects. Thus, if the image domain Ω satisfies

$$\Omega = A_1 \cup A_2 \cup \ldots A_n,$$

a disjoint union where the A_i denote the traces of homogeneous objects, then the change of light can be modelled as $g = (g_1, g_2, \ldots g_n)$, and the new image u_g is defined from the previous one v by

$$u_{|A_i} = g_i(v_{|A_i}). \tag{3}$$

The transparency phenomenon is not restricted to shadows or glass. It very frequently occurs with all kinds of spots on objects caused by more or less transparent liquids: water, painting, oil, mud, ink. Since we live in a more and more graphical world, with artificial light, we constantly see flat surfaces (paper, screen, walls) where occlusion and transparency processes have been repeatedly applied.

Let us mention two more important optical effects which increase still more the intricacy of natural images: first the shading, which makes elements of a homogeneous surface respond in a different way according to their orientation in space. The shading effect implies that a homogeneous object will be visually split into as many parts as facets it has, separated by "edges", that is discontinuity lines due to the sudden changes of orientation of the surface. These changes can also be gradual when the object's surface is smooth. Second phenomenon, the *reflectance* effect, or mirroring effect, which makes the surface of some object reflect the image of other objects or light sources. The extreme case for this phenomenon occurs with mirrors, but most objects present reflectance phenomena. In fact, shading may multiply the number of visual objects (as many as facets of homogeneous objects) and reflectance can be analyzed as a tranparency phenomenon. An extreme, but current effect of transparency occurs when we look through a glass window: the light coming from outsides mixes with the inside light reflected by the window. Calling respectively v_1, v_2 and u the inside image, the outside image and the resulting image, we obtain $u = g_1(v_1) + g_2(v_2)$. (Indeed, v_1 and v_2 are subject to an attenuation (contrast change) respectively due to reflection and transparency.) We may lead back the window effect to a transparency phenomenon : It is enough to consider each homogeneous region of v_1 as a spot S, inside which the image v_2 has been translated in grey level. This corresponds to an extension of our definition of transparency. In fact, the resulting image u can be defined as a "view of v_1 through the inhomogeneous transparent medium v_2", and conversely. The preceding model is compatible with a wide range of transparency phenomena. Transparency- or shadowing-like phenomena are produced in large variety of physical situations: a good example is the case of wetting, which can change the brightness of any objects by patches, in the same way as shadows do. This occurs (e.g.) on the ground when it is covered with puddles.

2.3. Requirements for image analysis operators. Of course, we do not know a priori, when we look at an image, what are the physical objects which have left a visual trace in it. We know, however, that the operations having led to the actual image are given by formulas (1)–(3). Thus, any processing of the image should avoid to destroy the image structure resulting from (1), (2) and (3): this structure traduces the information about shadows, facets and occlusions which are clues to the physical organization of the scene in space. *The identity and shape of objects must be recovered from the image by means which should be stable with respect to those operations.* As a basic and important example, let us recall how the Mathematical Morphology school has insisted on the fact that image analysis operations must be invariant with respect to any contrast change g. An image analysis operator T, acting on images u is called *morphological* if

$$T(g(u)) = g(T(u))$$

for any nondecreasing contrast change g. Classical examples of morphological operators are erosion, dilation, opening and closing.

Remark 2 (constrast change and response functions of captors)
One of the reasons for which actual contrast changes altering the "true" light intensity are nonlinear is following. Most light captors have a nonlinear behavior and, even worse, have a finite range. Whenever light is too strong (or too weak), saturation of the captor occurs. The response function of light captors generally has the shape showed in Figure 3.

3. Basic objects of image analysis

We call *basic objects* a class of mathematical objects, simpler to handle than the whole image, but into which any image can be decomposed and from which it can be reconstructed. The classical examples of image decompositions are

• Linear decompositions into simple waves: Basic objects of Fourier analysis are cosine and sine functions, basic objects of Wavelet analysis are wavelets or wavelet packets, basic objects of Gabor analysis are windowed (gaussian modulated) sines

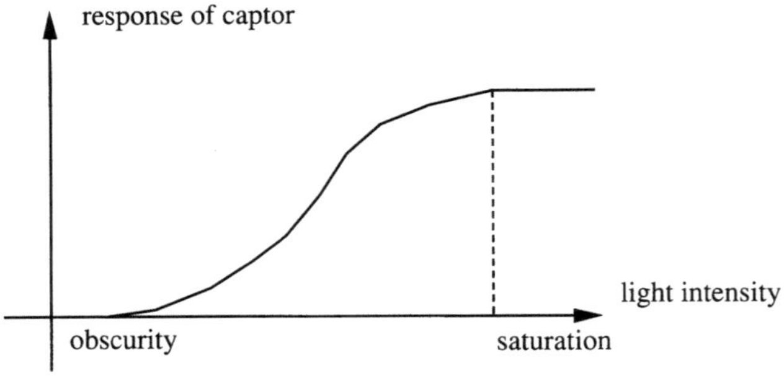

Figure 3: Nonlinear response of captors

and cosines. In all of these cases the decomposition is an additive one and we have argued against it as not adapted to the structure of images, except for restoration processes. Indeed, operations leading to the construction of real world images are basically nonlinear and the simplest of them, the constrast change, does not preserve linear decompositions. Indeed, if $u = u_1 + u_2$, then it is not true that $g(u) = g(u_1) + g(u_2)$ if the constrast change g is nonlinear.

• Next, we have the representation of the image by a *segmentation*, that is, a decomposition of the image into homogeneous regions separated by boundaries, or "edges". The criterion for the creation of edges or boundaries is the strength of constrast on the edges, along with the homogeneity of regions. Both of these criteria are simply not invariant with respect to contrast changes. Indeed $\nabla g(u) = g'(u)\nabla u$ so that we can alter the value of the gradient by simply changing the constrast.

• Finally, we can decompose, as proposed by the Mathematical Morphology school, an image into its binary shadows (or level sets), that is, we set $X_\lambda u(x) = white$ if $u(x) \geq \lambda$ and $X_\lambda u(x) = black$ otherwise. The white set is then called *level set* of u. It is easily seen that, an image can be reconstructed from its level sets by the formula

$$u(x) = \sup\{\lambda, u(x) \geq \lambda\} = \sup\{\lambda, x \in X_\lambda\}.$$

The decomposition therefore is nonlinear, and reduces the image to a family of plane sets (X_λ). In the discrete framework, all we can associate with each level set a finite set of Jordan curves which define its boundary. Conversely, the level set is uniquely defined from those Jordan curves. We shall call them *level curves of the image*. Are level sets and level curves the sought for basic objects of image processing? In some sense they are better than all above discussed "basic objects" because they are invariant under contrast changes. To be more precise, if we transform an image u into gou, where g is an increasing continuous function (understood as a contrast change), then it is easily seen that the set of level sets of gou is equal to the set of level sets of u.

However, level sets are drastically altered by occlusion or shadowing. Let us discuss this point. One can see in Figure 4 an elementary example of image generated by occlusion. A grey disk is partly occluded by a darker square (a). In (b) we display a perspective view of the image graph. In (c) and (d) we see two of the four distinct level sets of the image, the other ones being the empty set and the whole plane. It is easily seen that none of the level sets (c) and (d) corresponds to physical objects in the image. Indeed, one of them results from their union and the other one from their set difference. The same thing is true for the level lines (e) and (f): they appear as the result of some "cut and paste" process applied to the level lines of the original objects. It is nontheless true, because of the invariance with respect to constrast changes, that the basic objects of image processing must be somehow parts of the level lines. This leads us to what will be our proposition.

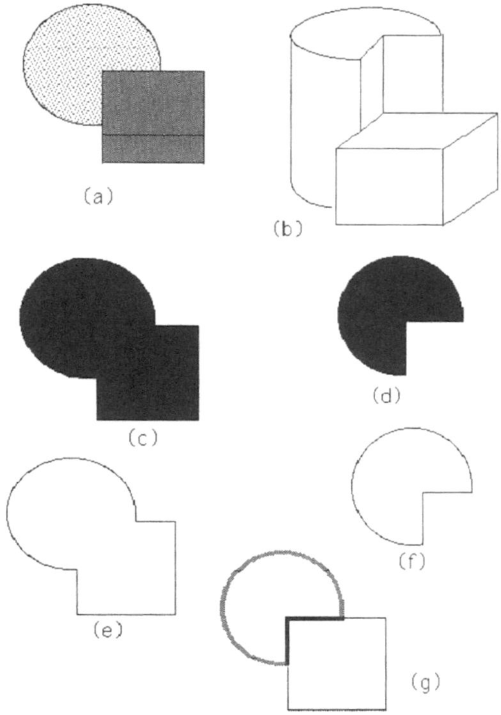

Figure 4: Image, level sets, level lines and T-junctions

- **Our proposition: Basic objects are T-junctions, X-junctions, Y-junctions and parts of level lines joining them.**

Following the phenomenological intuition of Kanizsa, we wish to define significant parts of images as the result of a segmentation of level lines. We refer to Figure 4 for a first simple (but, in fact, general) example. The level lines (e) and (f) represent two level lines at two different levels and in (g) we have superposed them in order to put in evidence how they are organized with respect to each other. We have displayed one of them as a thin line, the other one as a bold line and the common part in grey. The T-junctions can be defined in such an ideal case as the points where two level curves corresponding to two different levels meet and remain together.

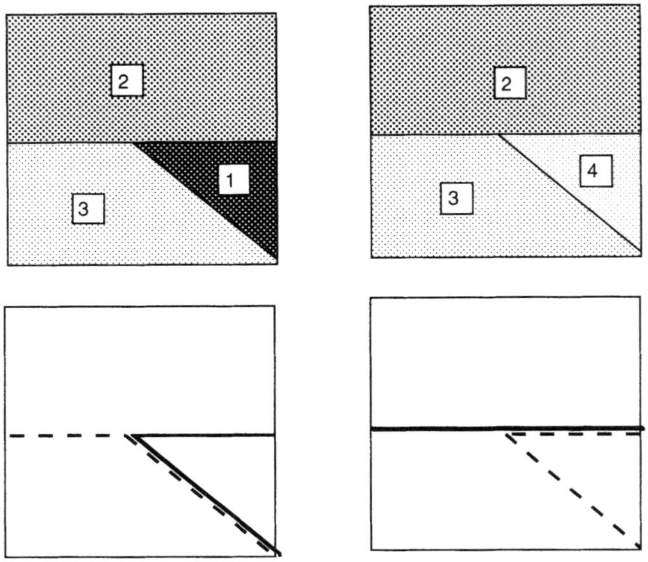

Figure 5: The two kinds of T-junctions depending on the ordering of grey levels

3.1. T-junctions. As shown in Figure 5, there are two possibilities which depend upon the order of grey levels around the T-junction. In the first case, the T-junction appears as two joint level lines taking at the junction two opposite directions. In the second case, one of the level lines is locally straight and the other one has two branches, one of which coincides with the first level line. Thus a T-junction at a point x can be (a little bit naively) described in the following way.

• Two level lines meet at point x.

• The half level lines (or branches) starting from x are thus four in number ; two coincide and two take opposite directions. (The branches which coincide may belong to different level lines or to the same. The branches which take opposite directions may also belong to the same level line or not).

The conditions on the orientation of the branches are absolutely necessary in order to distinguish T-junctions from other phenomena causing the meeting of level lines. First of all, the grey level quantization can provoke many meetings of level lines. Figure 6 shows this phenomenon for the grey level function $u(x, y) = -x-y$, where x and y denote cartesian coordinates after it has been digitized. In principle, the level lines of this function are parallel and never meet. Now, if the grid of discretization has a mesh equal to 1, then succesive and should-be-parallel level lines meet at every point of the grid. Another significant example of meeting point for level lines is the case where the image displays the meeting of three facets (and

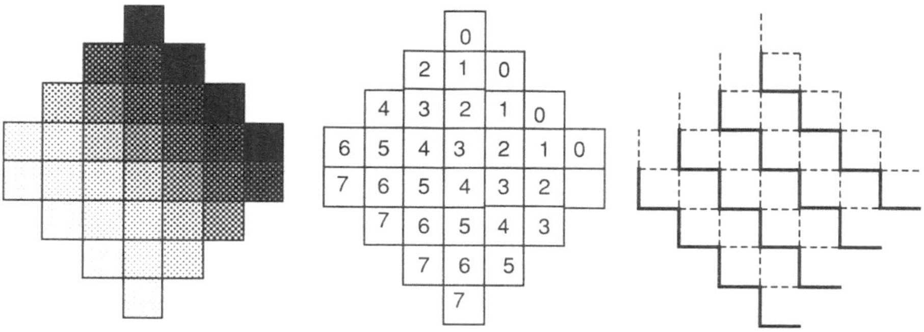

Figure 6: A "dégradé", its grey levels and its level lines. This illustrates the fact that the meeting of two level lines is no criterion for T-junction or "transparency".

Figure 7: Edges of a cube produce junctions of level lines which are not T-junctions, but rather "Y-junctions".

three edges) as displayed in Figure 7. In this case, we have a Y-junction, that is, no paralelism between branches can be observed.

Notice how the detection of T-junctions and their perceptual impact is a very local phenomenon, as is illustrated in Figure 8, where the addition of a tiny subjective boundary at the vertices of a rectangle creates the illusion of an overlapping. (We take this experiment from Kanizsa [Ka 2]).

3.2. Transparencies and X-junctions. The next case of singularity is caused by the transparency phenomenon and we shall call it "X-junctions". In the transparency phenomenon, we have assumed that inside a spot S, an original image v is changed into $u = g(v)$, where g is a contrast change. We keep $u = v$ outside the spot S. As a consequence, a level line with level λ outside S becomes a level line with level $g(\lambda)$ inside the spot, so that the level lines inside S are continuations of level lines outside S. Kanizsa showed that this results in the *continuation* perceptual phenomenon: we tend to see the level line of v crossing the boundary of S. Of course, this illusion is based on the continuity of direction. In the same way, of course, the boundary of S is seen as crossing the level lines of v. In fact, level lines cannot cross and the behaviour of level lines is analyzed in Figure 9. In the transparency phenomenon, the apparent crossing, which we shall call X-junction, consists in the meeting at a point x of two level lines. These level lines locally create four angular regions. Without loss of generality, we can assume that the angular regions have grey levels 1, 2, 3 and 4. Indeed, the grey levels need not

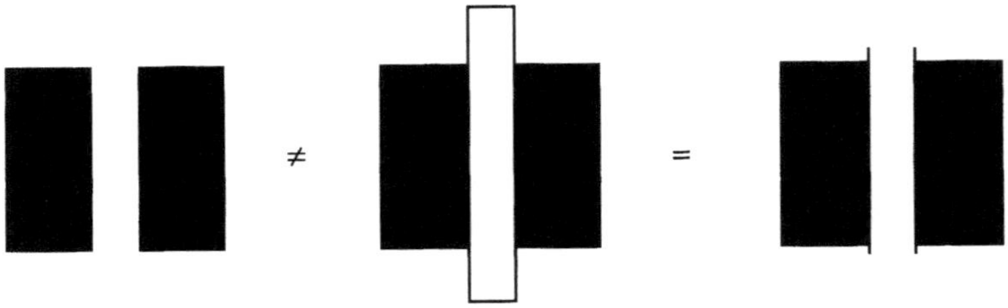

Figure 8: Locality of the T-junction effect: the left-hand figure is not interpreted as the result of the occlusion of a black rectangle by a white stripe while the right-figure is seen as equivalent to the central figure. This shows that tiny branches of "T" are enough to create a T-junction and produce a global occlusion effect.

be constant inside each region, but we know that the ranges of the four angular region are ordered. This is an obvious consequence of the fact that each pair of regions is separated by a level line. Then we see that three cases may occur, which are shown in Figure 9.

• First case: "checkerboard". The higher grey levels 3 and 4 are in two diagonally opposite regions and the lower grey levels 1 and 2 in the other two opposite regions. (The perfect checkerboard case corresponds to 3=4 and 1=2, but we display a more general case with the same structure for level lines). The checkerboard singularity does not seem to correspond to many physical phenomena, but is nontheless frequent in the human world, probably for the designer's pleasure of showing a parodoxal, physically impossible, visual situation. In this case, we can see how an intermediate level line forms a cross.

• Second case: "shadowing". The lower regions 1 and 2 are adjacent and so are the regions 3 and 4. In addition, 1 is adjacent to 3 and 2 to 4. Somehow, this situation is the most stable and least paradoxical of all, since it can be interpreted as the crossing of a shadow line with an edge. It is worth noticing that in this case the level line which separates the sets $\{x, u(x) > 2\}$ and $\{x, u(x) < 3\}$ can be interpreted indifferently as the shadow line or the edge line. The other (shadow or edge) line has no existence as a level line, but is obtained by concatenating two branches of two distinct level lines. The level lines bounding the sets $u > 1$ and $u > 3$ only meet at point x and form the "X-junction" which we shall trace in experiments on real images.

• Third case: "transparency". The regions 1 and 2 with lower grey-level and the higher grey-level regions 3 and 4 are again adjacent, but the pairs of extreme regions also are adjacent : 1 to 4 and 2 to 3. The transparent material is adding its own color, which is assumed here to be light grey, so that white becomes light grey and black becomes dark grey. The interpretation is not twofold in this case: the original edge (horizontal in the figure) must be the level line bounding the set $u > 2$. The shadow line (vertical in the figure), is built with two branches of distinct level lines.

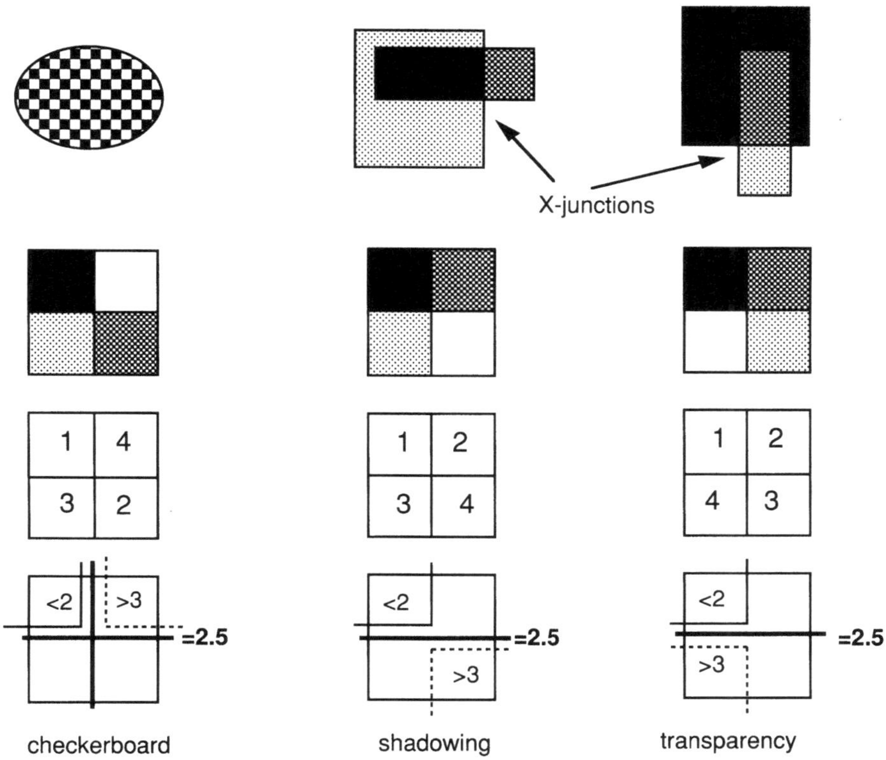

Figure 9: The three kinds of X-junctions. We first display (above) three kinds of visual experiments which lead respectively to checkerboard, shadowing and transparency sensations. The second and third rows focus on the resulting X-junctions and the local grey level values. The last row shows the resulting configuration of level lines. By "= 2.5", we denote the level lines separating the regions 1 and 2 from 3 and 4.

3.3. τ-junctions. This is the next and last case of singular meeting of level lines which we shall consider. We assume that two level lines meet at a point x and are tangent at this point. In addition, the level lines are disjoint on one side of x and coincide on the other side. We call this event a τ-junction at x. The perceptual (and physical) interpretation is as follows. An object with smooth surface is shown on a background with different colour. The object itself is not homogeneous and therefore has level lines inside. Because of the contrast with background, the apparent contour of the object coincides with level lines. Except for exceptional positions of the observer, the level lines inside the object's silhouette are seen to arrive tangently at the silhouette's boundary (Figure 10). Thus a typical event in our occluded world is following: a level line corresponding to the silhouette of an object meets tangent level lines on the one side (τ-junctions) and nontangent level lines (belonging to the background) on the other side. This second event creates T-junctions.

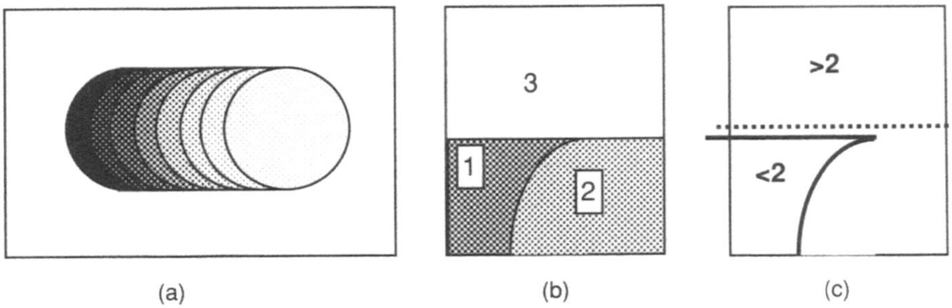

Figure 10: Tau-junctions. (a) physical situation where they occur, level lines on smooth object meeting the object's silhouette. (b) Local analysis of level sets. (c) Corresponding arrangement of level lines. The possible local arrangements of level lines are the same as for T-junctions.

It must be emphasized that we do not pretend to have completed the list of singularities of behaviour which can happen for level lines in a "natural" image. We have considered those singularities which seem to ask for a special treatment in the low level vision system, as has been shown by Kanizsa to be the case for T-junctions or X-junctions. It can be argued (and we have used such arguments) that the considered singularities are the most frequent in "natural" images of the physical world. There are, however, powerful arguments against this reduction. First of all, the fact that the physical world is by now full with industrial objects and synthetic documents of any kind. These objects can have any kind of shape and colour, so that no physical world argument remains valid for them. As an example, let us show how higher order singularities may be created in a very simple, physical and current visual event. If we place a solid cube on a plane surface in the sun light as displayed in Figure 11, we see how at a lower vertex of the cube a junction of four branches is created, like in the transparency phenomenon. Now, it is not a X-junction since no branches need be parallel or opposite. In short, it is a Y-junction (a vertex) with a shadow half line added. Thus, the meeting of two level lines at a point may have a physical interpretation without any angle condition. We can add this singularity to our classification of "natural visual events", but it is worth noticing that we can get more and more level lines at the same vertex by simply adding light sources: each new source adds a new shadow line at the same vertex. This event is of course very frequent with artificial lamps and can be seen in everynight's life on the street!

4. Experimental Kanizsa programme

4.1. Computation of basic visual events. In this section, we discuss how the basic perceptual-physical events discussed above can be detected in digital images and we present experimental results. The description of T-, X-, and τ-junctions is based on the assumptions that

- Level sets and level lines can be computed

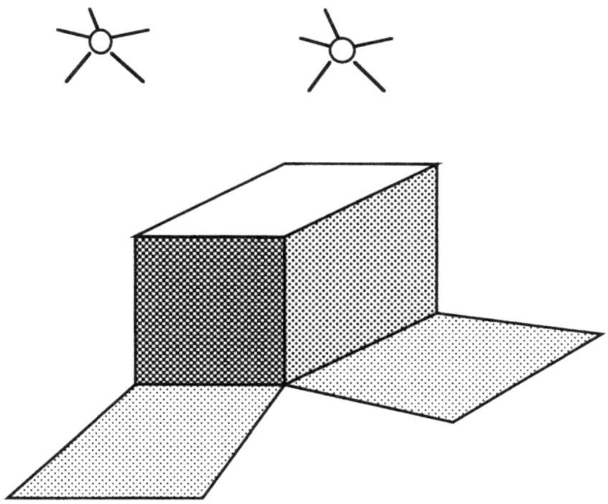

Figure 11: Higher order singularities. A cube with two light sources, two shadows and an order five singularity. This event is frequent with artificial light.

• The meeting of two level lines with different levels can be detected.

• The direction of level lines or level half lines can be defined and computed.

• The following events can be numerically implemented: "a branch (half level line) locally coincides with another", "a branch makes an angle with another", "two branches have opposite directions", "two branches are tangent".

Since in practice two different level lines in an image do not generally fully coincide, we see that all of these events involve the computation of a local direction for a branch of level line. In a digital image, the level sets are computed by simple thresholding. A level set $\{u(x) > \lambda\}$ is represented in white on black background. In the actual technology, $\lambda = 0, 1, \ldots, 255$, so that we can associate with an image 255 level sets. The Jordan curves limiting the level sets are easily computed by a straightforward algorithm: they are chains of vertical and horizontal segments limiting the pixels. In the numerical experiments, these chains are represented as chains of pixels by simply inserting" boundary pixels" between the actual image pixels.

We define "junctions" in general as every point of the image plane where two level lines (with different levels) meet. As we have commented in Figure 6, non significant T-junctions or X-junctions may be observed as a consequence of high gradient in the image. Assume that the image grid mesh is 1, with a grey level quantized with entire numbers. (In practice, from 0 to 255). Then at points where the gradient is larger than 1, collapsing of level lines may occur ; it necessarily occurs if the gradient is larger than $\sqrt{2}$. Thus all edges present in the image tend to generate T-junctions if the image is slightly blurry: indeed, the light smoothing of an edge generates a lot of parallel level lines which collapse at many points, thus generating tiny X-junctions. This fact can be appreciated in Experiment 2, where we have displayed all meetings of level lines in the image. As a consequence, the meeting of two level lines can be considered as physically meaningful if and only if the level lines diverge significantly from each other. In order to distinguish between the true junctions and the ones due tó quantization effects, we decide to take into account T-junctions if and only if first the area of the occulting object is large enough and second the apparent area of the occulted object is large enough too. This threshold on area must be of course as low as possible, because there is a risk to loose T-junctions between small objects. Small size significant objects appear very often in natural images: they may simply be objects seen at a large distance. In any case, the area threshold only depends on the quantization parameters of the digital image and tends ideally to zero when the accuracy of the image tends to infinite. The algorithm for the elimination of dubious junctions is as follows.

• Fix an area threshold n (in practice, $n = 40$ seems sufficient to eliminate junctions due to quantization effects.)

• At every point x where two level lines meet: define $\lambda_0 < \mu_0$ the minimum and maximum value of u in the four neighboring pixels of x.

• We denote by L_λ the connected component of x in the set $\{x, u(x) \leq \lambda\}$ and by M_μ the connected component of x in the set $\{x, u(x) \geq \mu\}$. Find the smallest $\lambda \geq \lambda_0$ such that the area of L_λ is larger than n. Call this value λ_1. Find the largest $\mu, \lambda_1 \leq \mu \leq \mu_0$, such that the area of M_μ is larger than n.

• If λ_1 and μ_1 have been found, then retain x as a valid junction.

Aknowledgement. We thank Marcella Mairota, Denis Pasquignon, Jacques Froment, Jean-Pierre d'Alès, Antonin Chambolle and Frédéric Guichard for valuable information and discussions.

Experiments

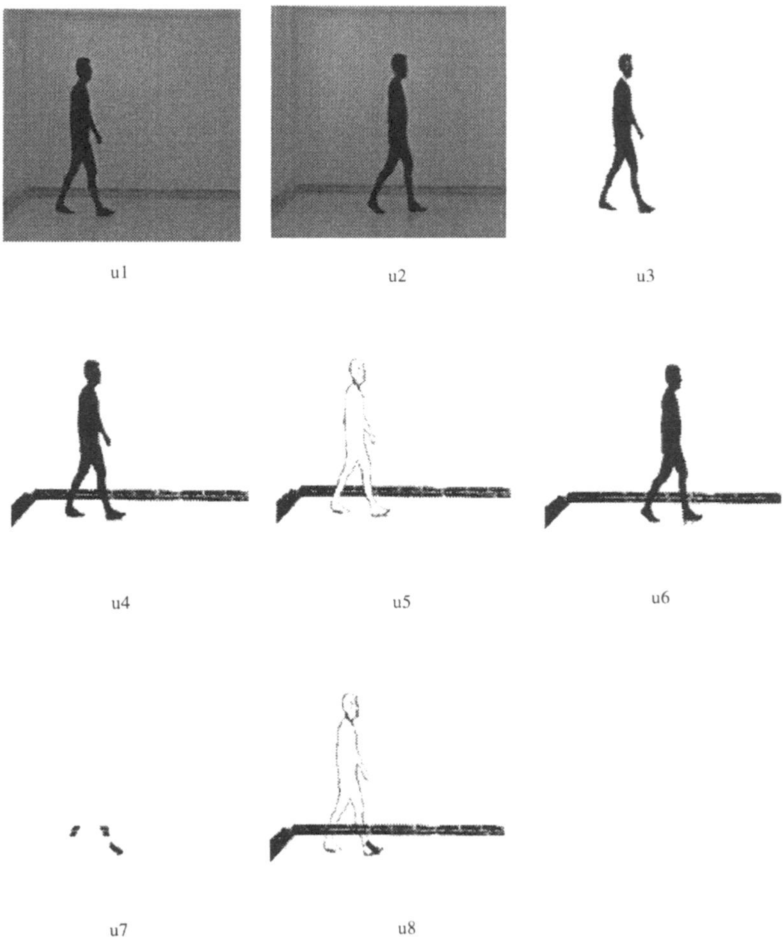

• **Experiment 1: The structure of occlusion retrieved by level set analysis in two successive images of a sequence**. The eight images $u1, \dots, u8$ of the experiment are displayed from left to right and from top to bottom. The images $u1$ and $u2$ are grey-level images and the other ones are binary images (black and white) which display level sets and logical operations effectuated on them. The coding convention is $black = 0 = false$, $white = 255 = true$, so that we identify level sets with binary images and union with max, intersection with min and set difference with "−". We first see *el caminante* (a walking researcher, photographed by Paco Perales) in two successive positions $u1$ and $u2$. Then *el caminante* without skirting-board: $u3 = \{x, u1(x) > 50\}$ and *el caminante* with skirting-board $u4 = \{x, u1(x) > 140\}$. Next, $u5 = u4 - u3$, that is, the skirting-board without the walking man, $u6 = \{x, u2(x) > 140\}$, *El caminante* with skirting-board in Position 2. Finally $u7 = max(u3, u6)$: the resulting black parts display the part of the skirting-board occluded in Position 1 plus the part of the ground which remains occluded by the feet of the *caminante* in both images. In $u8 = min(u7, u5)$, we display the reconstruction of the whole skirting-board by adding the occluded part in Position 1 deduced from Position 2.

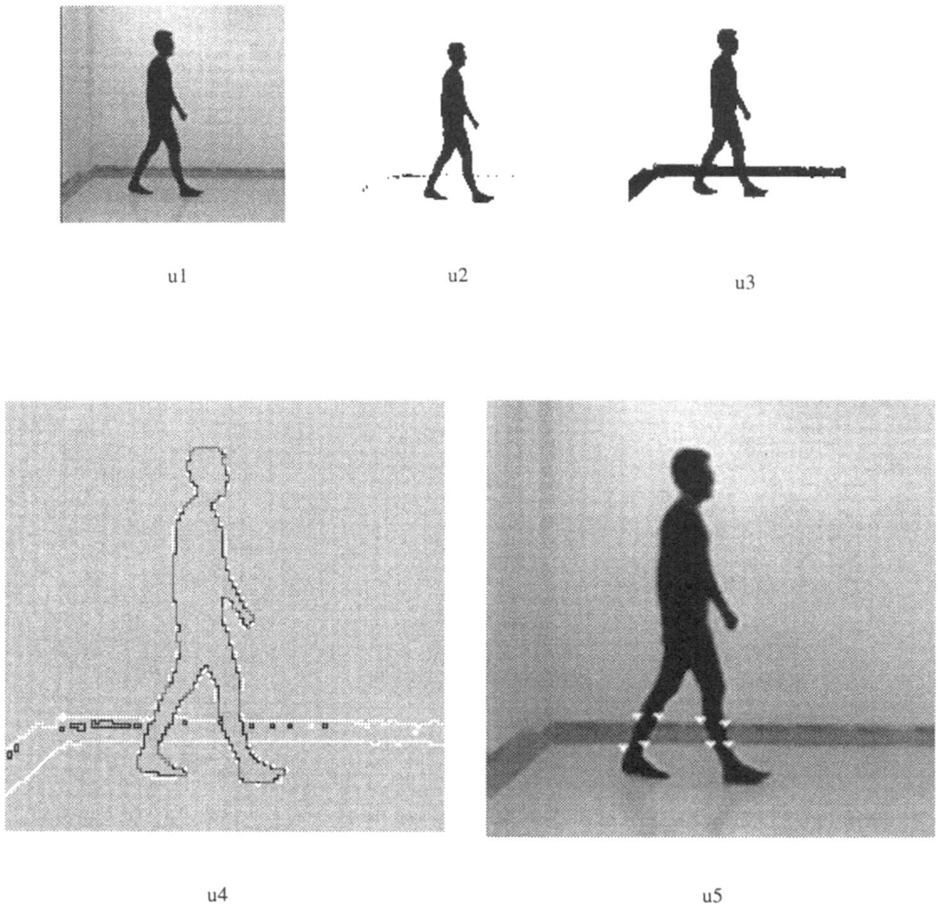

• **Experiment 2: detection of T-junctions as meeting of level lines.** The original grey-level image is $u1$. In $u2$ we see *El caminante* without skirting-board: $u2 = \{u1 > 90\}$ and $u3 = \{u1 > 140\}$ is *El caminante* with skirting-board. We display in $u4$ the level lines of $u2$ in white and the level lines of $u3$ in black with their common parts in dark grey. Candidates to T-junctions generated by the main occlusions can be seen. They are characterized as meeting points of a black, a white and a grey line. In $u5$, we see the detected T-junctions after a filtering of spurious T-junctions has been made, with area threshold 12.

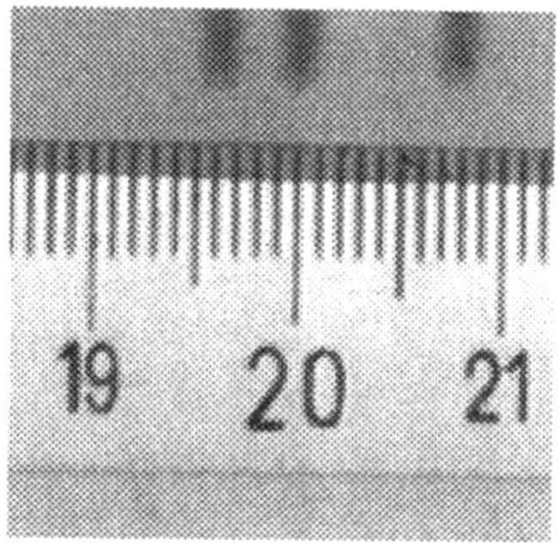

u1

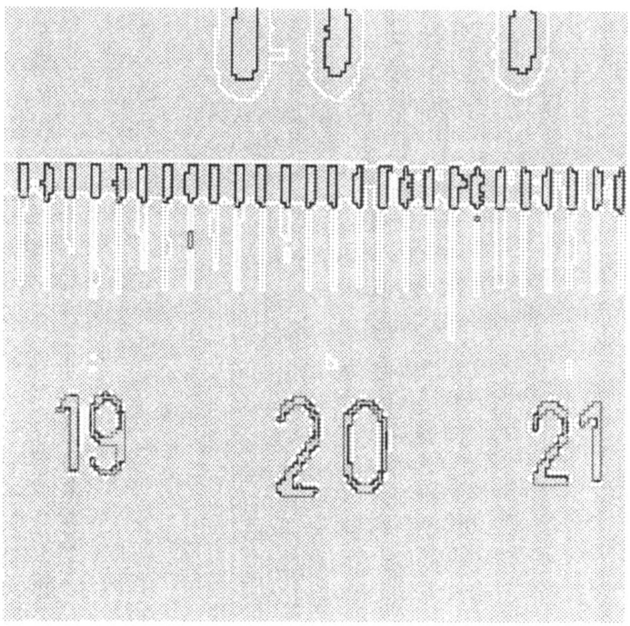

u2

• **Experiment 3: Transparencies** The criterion for transparency is: a meeting point of level lines such that the angle between two disjoint pairs of branches is close to π. The original image $u1$ displays a transparent plastic ruler superposed to a white paper on the border of a biological experiment. The next image shows the level lines at two well chosen levels and the characteristic X-junctions generated by the transparency phenomenon, as predicted.

u1

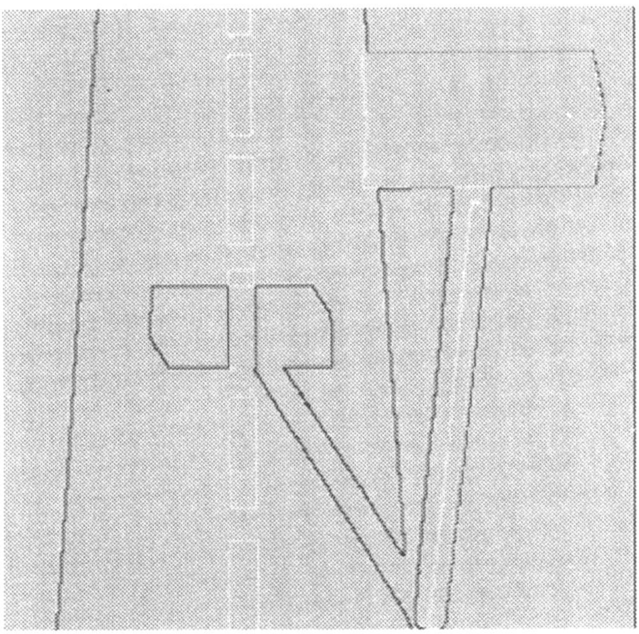

u2

• **Experiment 4: Transparencies and T-junctions provoked by shadows in a synthetic image.** The image $u1$ is a synthetic image with rendering effects (author Joan Montes de Oca). In $u2$ we display, with the same convention as in Experiments 2 and 3, the level lines at two well chosen levels. One can see a X-junction on the road, due to a shadow and several T-junctions due to occluding contours. There also is a $\tau - junction$, caused by a shadow contour on a cylinder meeting the apparent contour of the cylinder.

References

[A] J. Alison Noble. Finding half boundaries and junctions in images. Image and Vision Computing, vol.10, 4, may 1992.

[AM] L. Alvarez and Freya Morales. Affine Morphological Multiscale Analysis of Corners and Multiple Junctions. Ref 9402, August 1994, Dept. Informatica y Systemas, Universidad de Las Palmas de Gran Canaria.

[Ar] R. Arnheim. *Art and visual perception: a psychology of the creative eye*, 1954.

[Ar] R. Arnheim. *Arte e percezione visiva*, Feltrinelli, Milano, (tenth edition), 1993.

[DG] R. Deriche and G. Giraudon. A computational approach for corner and vertex detection, International Journal of Computer Vision, 10, 2. 101–124, 1993.

[Ka1] G. Kanizsa. *Organization in Vision*, N.Y. Praeger, 1979.

[Ka2] G. Kanisza. *Grammatica del Vedere*, Il Mulino, Bologna, 1980.

[Ka3] G. Kanisza *Vedere e pensare*, Il Mulino, Bologna, 1991.

[Li] T. Lindeberg. Junction detection with automatic selection of detection scales and localization scales in Proceedings of ICIP-94, Texas 1994.

[Ma] D. Marr. *Vision* N.York, W.H. Freeman and Co, 1982.

[Me] Y. Meyer. Les ondelettes, algorithmes et applications, Armand Colin, Paris, 1992.

[MS] J.M. Morel and S. Solimini. Variational methods in image processing, Birkhäuser, 1994.

[NMS] M. Nitzberg, D.Mumford and T. Shiota. Filtering, Segmentation and Depth. Lecture Notes in Computer Science, 662, Springer-Verlag 1993.

[NM] M. Nitzberg and D. Mumford. "The 2.1 Sketch", in Proceedings of the Third International Conference on Computer Vision, Osaka 1990.

[We] M. Wertheimer. Untersuchungen zur Lehre der Gestalt, II, Psychologische Forschung, vol. 4, pp.301–350, 1923.

[Wi] L. Williams. Perceptual organization of occluding contours, Proceedings of the third International Conference on Computer Vision, Osaka 1990.

Progress in Nonlinear Differential Equations
and Their Applications, Vol. 25

A Second Order Model in Image Segmentation: Blake & Zisserman Functional

MICHELE CARRIERO* ANTONIO LEACI* FRANCO TOMARELLI⋆

* Dipartimento di Matematica, Via Arnesano, 73100 Lecce, Italia
⋆ Dipartimento di Matematica, Politecnico, Piazza Leonardo da Vinci 32, 20133 Milano, Italia

ABSTRACT. We state a weak formulation for the Blake & Zissermann model of image segmentation and prove the existence of a minimizer for such a problem.

1. Introduction

In recent years many variational principles related to image segmentation have been introduced and studied: an updated survey on the subject can be found in [MSo], [R].

About the first order functional introduced by [MSh] (the weak membrane model)

$$\int_{\Omega\setminus K} \left(|Du|^2 + \mu|u-g|^2\right) dy + \alpha\mathcal{H}^{n-1}(K\cap\Omega) \quad \text{to be minimized over closed sets } K \text{ and } u \in C^1(\Omega\setminus K) \tag{1.1}$$

many deep theoretical results are already available ([DA], [DGCL], [DMS], [AFP], [DS], [LS]), though some important questions are left open.

About the second order functional introduced by [BZ] (the thin plate surface under tension)

$$\begin{aligned} F(K_0,K_1,u) = &\int_{\Omega\setminus(K_0\cup K_1)} \left(|D^2u|^2 + \mu|u-g|^2\right) dy + \\ &+ \alpha\mathcal{H}^{n-1}(K_0\cap\Omega) + \beta\mathcal{H}^{n-1}((K_1\setminus K_0)\cap\Omega) \\ &\text{to be minimized over closed sets } K_0, K_1 \text{ and} \\ &u\in C^0(\Omega\setminus K_0)\cap C^2(\Omega\setminus(K_0\cup K_1)) \end{aligned} \tag{1.2}$$

(where K_0 is the set of jump points for u and $K_1 \setminus K_0$ is the set of crease points) almost no theoretical statement is proved besides some convergence for a numerical approximation scheme [BZ], and the one-dimensional model problem [C], [B].

In this paper we succeed in formulating and solve a weak version of the n-dimensional problem. More precisely we introduce the following weak formulation of problem (1.2), (see Definition 2)

$$\boxed{\begin{aligned}\mathcal{F}(u) = \int_\Omega (|\nabla^2 u|^2 + \mu|u-g|^2)\,dy\, + \\ + \alpha\mathcal{H}^{n-1}(S_u) + \beta\mathcal{H}^{n-1}(S_{\nabla u}\setminus S_u) \\ \text{to be minimized over } u \in GSBV(\Omega), \text{ with } \nabla u \in [GSBV(\Omega)]^n\end{aligned}} \tag{1.3}$$

and we prove the existence of a minimizer for such a problem in any space dimension n. In (1.3) ∇u denotes the absolutely continuous part of the distributional derivative Du.

The main result (Theorem 1) was announced in [CLT4]. The proof relies on the direct method of the Calculus of Variations and is splitted in the following steps: slicing properties of first and second derivatives (Theorems 2, 3), an interpolation inequality (Theorems 5, 6), a compactness theorem for the sublevels of the functional $\mathcal{F}$ (Theorem 8) and a lower semicontinuity property with respect to a.e. convergence (Theorem 10).

The substantial difficulty when looking for a finite energy space fulfilling compactness and lower-semicontinuity of minimizing sequences is that finiteness of (1.3) does not provide any estimate on Du or even on ∇u. More precisely $|\nabla u|$ may be essentially unbounded even if the energy (1.3) is arbitrarily small, as shown by the example (1.4). In addition, the fact that the competing functions u are allowed to be discontinuous immediately drives the analysis outside the framework of the space of functions with Special Bounded Hessian $SBH(\Omega)$ ([CLT1]) where second order energies for elastic-perfectly plastic plates and rigid-plastic slabs have been successfully studied [CLT1,2,4], [T].

Moreover, truncating the competing functions is allowed in the finite energy space, but truncation jeopardizes the estimates on D^2u since (1.3) does not provide any information on the jump of ∇u, in contrast to the Lipschitz behaviour of truncation in $BH(\Omega)$ (see [CLT3] and [S]). On the other hand, by allowing discontinuous competing functions u belonging to the space of functions with Special Bounded Variation $SBV(\Omega)$ such that the approximate gradient ∇u belongs to $[SBV(\Omega)]^n$, one loses the compactness of the sublevels of the functional $\mathcal{F}$ even with respect to the a.e. convergence, as shown by the following example. For $y=(y_1,y_2)\in\mathbf{R}^2$ and $h\in\mathbf{N}, h\geq 4$ set $r_h = 2^{-h-1}$ and

$$v_h(y_1,y_2) = \begin{cases} \frac{4^k}{k}\left(y_1 - \frac{1}{k}\right) & \text{if } y\in B_{r_k}\left(\left(\frac{1}{k},0\right)\right), \quad k=4,\dots,h \\ 0 & \text{if } y \notin \cup_{k=4}^h B_{r_k}\left(\left(\frac{1}{k},0\right)\right). \end{cases} \tag{1.4}$$

We have

$$\mathcal{F}(v_h) = \mu\int_{\mathbf{R}^2} |v_h|^2 dy + \alpha\mathcal{H}^1(S_{v_h}) = \mu\frac{\pi}{64}\sum_{k=4}^h \frac{1}{k^2} + \alpha\pi\sum_{k=4}^h \frac{1}{2^k} \leq c < +\infty.$$

The sequence converges in $L^2(\mathbf{R}^2)$ to the function v_0 given by

$$v_0(y) = v_0(y_1, y_2) = \begin{cases} \frac{4^k}{k}\left(y_1 - \frac{1}{k}\right) & \text{if } (y_1, y_2) \in B_{r_k}\left(\left(\frac{1}{k}, 0\right)\right), \ k = 4, \dots \\ 0 & \text{if } (y_1, y_2) \notin \bigcup_{k=4}^{\infty} B_{r_k}\left(\left(\frac{1}{k}, 0\right)\right). \end{cases}$$

We see that ∇v_0 is not $L^1_{\mathrm{loc}}(\mathbf{R}^2)$, hence $v_0 \notin SBV^2_{\mathrm{loc}}(\mathbf{R}^2)$ and even $v_0 \notin SBV_{\mathrm{loc}}(\mathbf{R}^2)$.

Besides the application to segmentation theory, the techniques used in this paper are of interest in as much as (1.3) is an example of a functional in the calculus of variations whose computation (as in the case of perimeter) is not reducible to one-dimensional slices; however, $\mathcal{F}$ can be estimated through lower dimensional sections; in this way we deal with mixed derivatives in paragraph 6.

2. Notations and Main Result

From now on we assume $n \geq 2$ and that $\Omega \subset \mathbf{R}^n$ is an open set, μ, α, β are positive parameters and $g \in L^2(\Omega)$.

We recall that in the Blake & Zisserman model μ is a scale parameter, α is a contrast parameter and a measure of immunity to noise, β is a gradient contrast parameter.

For a given set $U \subset \mathbf{R}^n$ we denote by ∂U its topological boundary, by $\mathcal{H}^k(U)$ its k-dimensional Hausdorff measure and by $|U|$ its Lebesgue outer measure, and χ_U is the characteristic function of U. We indicate by $B_\rho(x)$ the open ball $\{y \in \mathbf{R}^n; |y - x| < \rho\}$, and we set $B_\rho = B_\rho(0)$. If Ω, Ω' are open subsets in $\mathbf{R}^n$, by $\Omega \subset\subset \Omega'$ we mean that $\overline{\Omega}$ is compact and $\overline{\Omega} \subset \Omega'$. We introduce the following notations: $\alpha \wedge \beta = \min\{\alpha, \beta\}$, $\alpha \vee \beta = \max\{\alpha, \beta\}$ for every $\alpha, \beta \in \mathbf{R}$; given two vectors a, b, we set $a \cdot b = \sum_i a_i b_i$.

For any Borel function $v : \Omega \to \mathbf{R}$ the approximate upper and lower limits of v are the Borel functions $v^+, v^- : \Omega \to \bar{\mathbf{R}} = \mathbf{R} \cup \{\pm\infty\}$ defined for any $x \in \Omega$ by

$$v^+(x) = \inf\{t \in \bar{\mathbf{R}} : \lim_{\rho \to 0} \rho^{-n} |\{u > t\} \cap B_\rho(x)| = 0\},$$
$$v^-(x) = \sup\{t \in \bar{\mathbf{R}} : \lim_{\rho \to 0} \rho^{-n} |\{u < t\} \cap B_\rho(x)| = 0\}.$$

The set

$$S_v = \{x \in \Omega; \ v^-(x) < v^+(x) \}$$

is a Borel set of negligible Lebesgue measure (see e.g. [F], 2.9.13); we say that the approximate limit of v exists in $\Omega \setminus S_v$ and we denote by $\tilde{v} : \Omega \setminus S_v \to \bar{\mathbf{R}}$ the function

$$\tilde{v}(x) = \operatorname{ap}\lim_{y \to x} v(y) = v^+(x) = v^-(x).$$

Let $x \in \Omega \setminus S_v$ be such that $\tilde{v}(x) \in \mathbf{R}$; we say that v is approximately differentiable at x if there exists a vector $\nabla v(x) \in \mathbf{R}^n$ such that

$$\operatorname{ap}\lim_{y \to x} \frac{|v(y) - \tilde{v}(x) - \nabla v(x) \cdot (y - x)|}{|y - x|} = 0.$$

If v is a smooth function then ∇v is the gradient. In the following by the notation $|\nabla v|$ we mean the euclidean norm of ∇v and we set $\nabla_i v = (\mathbf{e}_i \cdot \nabla)v$, $\{\mathbf{e}_i\}$ denoting the canonical base of $\mathbf{R}^n$. In the one-dimensional case ($n = 1$) we shall use the notation v' in place of ∇v.

We recall the definition of the space of functions of bounded variation in Ω with values in $\mathbf{R}$:

$$BV(\Omega) = \{v \in L^1(\Omega); Dv \in \mathcal{M}_n(\Omega)\}$$

where $Dv = \{D_i v\}_{i=1,..n}$ denotes the partial distributional derivatives of v and $\mathcal{M}_n(\Omega)$ denotes the space of vector-valued Radon measures with finite total variation. We denote by $|\cdot|_T$ the total variation of a measure.

For every $v \in BV(\Omega)$ the following properties hold:

1) $\widetilde{v}(x) \in \mathbf{R}$ for $\mathcal{H}^{n-1}$-almost all $x \in \Omega \setminus S_v$ (see [F], 4.5.9(3));
2) S_v is countably $(\mathcal{H}^{n-1}; n-1)$ rectifiable (see [F], 4.5.9(16));
3) ∇v exists a.e. in Ω and coincides with the Radon-Nikodym derivative of Dv with respect to the Lebesgue measure (see [F], 4.5.9(26));
4) for $\mathcal{H}^{n-1}$ almost all $x \in S_v$ it holds $v^+(x), v^-(x) \in \mathbf{R}$ and there exists a unique $\nu = \nu_v(x) \in \partial B_1$ such that (see [Z], 5.14.3)

$$\lim_{\rho \to 0} \rho^{-n} \Big[\int_{\{y \in B_\rho(x); y\cdot\nu > 0\}} |v(y) - v^+(x)|\, dy + \\ + \int_{\{y \in B_\rho(x); y\cdot\nu < 0\}} |v(y) - v^-(x)|\, dy \Big] = 0,$$

and also (see [F], 4.5.9(15))

$$|Dv|_T \geq \int_\Omega |\nabla v|\, dy + \int_{S_v} |v^+ - v^-|\, d\mathcal{H}^{n-1}.$$

Moreover $\nu_v(x)$ is an approximate normal vector to S_v at x (see [Z], 5.9.6).

We recall the definitions of some function spaces and we refer to [DGA] and [A1] for their properties.

DEFINITION 1 *$SBV(\Omega)$ denotes the class of functions $v \in BV(\Omega)$ such that*

$$|Dv|_T = \int_\Omega |\nabla v|\, dy + \int_{S_v} |v^+ - v^-|\, d\mathcal{H}^{n-1}.$$

$SBV_{\text{loc}}(\Omega)$ denotes the class of functions $v \in SBV(\Omega')$ for every $\Omega' \subset\subset \Omega$.

DEFINITION 2 *For an open set $\Omega \subset \mathbf{R}^n$*

$$GBV(\Omega) = \big\{v : \Omega \to \mathbf{R} \text{ Borel function}; -k \vee v \wedge k \in BV_{\text{loc}}(\Omega)\ \forall k \in \mathbf{N}\big\},$$
$$GSBV(\Omega) = \big\{v : \Omega \to \mathbf{R} \text{ Borel function}; -k \vee v \wedge k \in SBV_{\text{loc}}(\Omega)\ \forall k \in \mathbf{N}\big\}.$$

LEMMA 1 *Let $v \in GSBV(\Omega)$. Then*
1) S_v *is countably* $(\mathcal{H}^{n-1}, n-1)$ *rectifiable;*
2) ∇v *exists a.e. in* Ω.

Proof See [A2], proposition 1.3 and 1.4.

Definition 2 has a very simple interpretation in dimension one, as is shown in the following lemma (whose proof is trivial); it shows that in dimension one strong formulation (1.2) and weak formulation (1.3) coincide and this will be used in the proof of the interpolation theorem.

LEMMA 2 *Let $a, b \in \mathbf{R}$ and $v \in GSBV(a,b)$. If $\int_a^b |v'(t)|dt + \mathcal{H}^0(S_v) < +\infty$ then $v \in SBV(a,b)$. If $\int_a^b |v'(t)|^2 dt < +\infty$ and $\mathcal{H}^0(S_v) = 0$ then $v \in H^{1,2}(a,b)$.*

Now we define the following function spaces; even if we will use only $GSBV^2(\Omega)$ in the following, in order to avoid confusion with other authors' notations we recall the definitions of various spaces related to bounded second derivatives.

DEFINITION 3 *For $\Omega \subset \mathbf{R}^n$ open set*

$$SBH(\Omega) = \{v \in W^{1,1}(\Omega),\ Dv \in [SBV(\Omega)]^n\}$$
$$SBV^2(\Omega) = \{v \in SBV(\Omega),\ \nabla v \in [SBV(\Omega)]^n\}$$
$$GSBV^2(\Omega) = \{v \in GSBV(\Omega),\ \nabla v \in [GSBV(\Omega)]^n\}$$

For functions in $SBH(\Omega)$, $SBV^2(\Omega)$ or $GSBV^2(\Omega)$ we use the notation $\nabla^2_{ij} v = \nabla_j(\nabla_i v)$, $\nabla^2_i v = \nabla_i(\nabla_i v)$ and, in the one-dimensional case, $v'' = (v')'$. Notice that $Dv = \nabla v$ in $SBH(\Omega)$, $Dv \neq \nabla v$ in $SBV^2(\Omega), GSBV^2(\Omega)$. Moreover we set

$$S_{\nabla v} = \cup_{i=1}^n S_{\nabla_i v}.$$

Now we introduce the weak energy functional and a space for competing functions.

DEFINITION 4 *For $\Omega \subset \mathbf{R}^n$ an open set, setting $X(\Omega) = GSBV^2(\Omega) \cap L^2(\Omega)$, we define*

$$\mathcal{F} : X(\Omega) \to [0, +\infty]$$

$$\mathcal{F}(v) = \int_\Omega (|\nabla^2 v|^2 + \mu|v-g|^2)\, dy + \alpha \mathcal{H}^{n-1}(S_v) + \beta \mathcal{H}^{n-1}(S_{\nabla v} \setminus S_v).$$

We need a localization of the functional $\mathcal{F}$. For every open set $A \subset \Omega$ and for every $v \in GSBV^2(\Omega)$ we set

$$\mathcal{F}(v, A) = \int_A \left(|\nabla^2 v|^2 + \mu|v-g|^2\right)\, dy + \alpha \mathcal{H}^{n-1}(S_v \cap A) + \beta \mathcal{H}^{n-1}((S_{\nabla v} \setminus S_v) \cap A).$$

Moreover for every open set $A \subset \mathbf{R}$ and for every $v \in GSBV^2(A)$ we set

$$\mathcal{F}_1(v, A) = \int_A \left(|v''|^2 + \mu|v - g|^2\right) dt + \alpha\mathcal{H}^0(S_v \cap A) + \beta\mathcal{H}^0((S_{v'} \setminus S_v) \cap A).$$

We state the main result whose proof relies on results of sections 3, 4, 5 and 6.

THEOREM 1 *Let $\Omega \subset \mathbf{R}^n$ be an open set. Assume $g \in L^2(\Omega)$, $0 < \mu$, $0 < \beta \leq \alpha \leq 2\beta$. Then there is $v_0 \in X(\Omega)$ such that*

$$\mathcal{F}(v_0) \leq \mathcal{F}(v) \qquad \forall v \in X(\Omega).$$

Proof If we choose $v \equiv 0$, then $\mathcal{F}(v) = \mu \int_\Omega g^2\, dy < +\infty$. Let $(v_h) \subset X(\Omega)$ be a minimizing sequence for $\mathcal{F}$. By Theorem 8 and a diagonalization argument we find a subsequence (v_{h_m}) and $v_0 \in X(\Omega)$ such that $v_{h_m} \to v_0$ a.e. in Ω. By Theorem 10, for every $k \in \mathbf{N}$

$$\mathcal{F}(v_0, \Omega \cap B_k) \leq \liminf_m \mathcal{F}(v_{h_m}, \Omega \cap B_k) \leq \liminf_m \mathcal{F}(v_{h_m}).$$

Taking the limit as $k \to +\infty$ we obtain

$$\mathcal{F}(v_0) = \inf_{v \in X(\Omega)} \mathcal{F}(v). \qquad \square$$

3. Slicing properties

In this section, for any fixed direction in $\mathbf{R}^n$, we define the fibers of Ω along such direction and study the traces on the fibres of functions defined on the whole Ω.

Let $\nu \in \partial B_1$ and define the orthogonal projection

$$\pi_\nu : \mathbf{R}^n \to \mathbf{R}^n, \qquad \pi_\nu(y) = y - (y \cdot \nu)\nu.$$

For every subset $E \subset \mathbf{R}^n$ we define, for every $x \in \pi_\nu(E)$,

$$E_x^\nu = \{t \in \mathbf{R};\ x + t\nu \in E\},$$

for every $v \in GSBV(\Omega)$ we define for $\mathcal{H}^{n-1}$-a.e. $x \in \pi_\nu(E)$,

$$v_x^\nu(t) = v(x + t\nu) \quad \text{(defined for a.e. } t \in E_x^\nu).$$

We prove some slicing properties in the case $\nu = \mathbf{e}_i$ and $\Omega = (0,1)^n$; for brevity's sake in this case we set $\pi_i = \pi_{\mathbf{e}_i}$ and we omit the label ν in the previous definitions.

THEOREM 2 *Let $\Omega = (0,1)^n$ and $v \in GBV(\Omega)$. Then for $\mathcal{H}^{n-1}$-a.e. $x \in \pi_i(\Omega)$ $(i = 1, \ldots, n)$*

(i) $v_x(\cdot) \in GBV(0,1)$,

(ii) $v_x{}'(t) = \nabla_i v(x + t\mathbf{e}_i)$ for a.e. $t \in (0,1)$,

(iii) $S_{v_x} = (S_v)_x$.

Moreover
(iv) $v \in GSBV(\Omega)$ *iff* $v_x \in GSBV(0,1)$ *for* $\mathcal{H}^{n-1}$-a.e. $x \in \pi_i(\Omega), \forall i$,
(v) *if* $v \in SBV(\Omega)$ *then* $v_x \in SBV(0,1)$ *for* $\mathcal{H}^{n-1}$-a.e. $x \in \pi_i(\Omega), \forall i$.

Proof The five properties follow from [A1], Theorem 3.3 and [A2], section 1.

THEOREM 3 *Let* $\Omega = (0,1)^n$ *and* $v \in GSBV^2(\Omega)$. *Then for* $\mathcal{H}^{n-1}$-a.e. $x \in \pi_i(\Omega)$ $(i = 1, \dots, n)$
(i) $v_x''(t) = \nabla_i^2 v(x + t\mathbf{e}_i)$ a.e. $t \in (0,1)$,
(ii) $S_{v_x'} \subseteq (S_{\nabla v})_x$,
(iii) $(v_x')^+(t) = (\nabla_i v)^+(x + t\mathbf{e}_i)$, $(v_x')^-(t) = (\nabla_i v)^-(x + t\mathbf{e}_i)$ $\forall t \in (0,1)$.

Proof The three properties follow from [A1] applied to $\nabla_i v$ and from the fact that

$$S_{\nabla v} = \cup_{i=1}^n S_{\nabla_i v},$$

hence for $\mathcal{H}^{n-1}$-a.e. $x \in \pi_i(\Omega)$

$$S_{v_x'} = \{t \in (0,1) : (x + t\mathbf{e}_i) \in S_{\nabla_i v}\} \subseteq \{t \in (0,1) : (x + t\mathbf{e}_i) \in S_{\nabla v}\}. \qquad \square$$

THEOREM 4 *For every* $\mathcal{H}^{n-1}$ *measurable set* $E \subset \mathbf{R}^n$, $(\mathcal{H}^{n-1}, n-1)$ *rectifiable and for every* $\nu \in \partial B_1$,

$$\int_{\pi_\nu(E)} \mathcal{H}^0(E_x)\, d\mathcal{H}^{n-1}(x) = \int_E |\nu \cdot \nu_E(y)| d\mathcal{H}^{n-1}(y) \le \mathcal{H}^{n-1}(E),$$

where $\nu_E(y)$ *is an approximate unit normal to* E *at* y.

Proof See [F], 3.2.22.

REMARK The definition of $GSBV^2(\Omega)$ and the properties proved in Theorems 2, 3 are invariant under translations and rotations of the coordinates (see [F], 3.1.4), hence the results obtained hold for sections of v in any direction.

4. Interpolation inequality

In the following we denote by (φ_k) a sequence of functions such that

$$\begin{cases} \varphi_k \in C^2(\mathbf{R}), & 0 \le \varphi_k' \le 1, \\ \varphi_k(t) = t & \forall t \in [-k+1, k-1] \\ |\varphi_k(t)| = k & \forall |t| > k+1. \end{cases} \tag{4.1}$$

REMARK The property v belongs to $GSBV(\Omega)$ (see Definition 2) is clearly equivalent to requiring that

$$\varphi_k \circ v \in SBV_{\mathrm{loc}}(\Omega) \quad \forall k \in \mathbf{N}$$

(see [A2], section 1).

THEOREM 5 *Assume* $a, b \in \mathbf{R}$, $k \in \mathbf{N}, v \in GSBV^2(a,b)$. *Then*

$$\int_a^b |(\varphi_k \circ v)'| dt \le 2k(\mathcal{H}^0(S_v \cup S_{v'}) + 1) + \frac{2}{3}(b-a)^{\frac{3}{2}} \left(\int_a^b |v''|^2 dt \right)^{\frac{1}{2}}.$$

Proof We may assume that the right hand side is finite; we set

$$t_0 = a,\ t_{m+1} = b \ \text{ and } \ S_v \cup S_{v'} = \bigcup_{i=1}^{m} \{t_i\},$$

with $t_{i-1} < t_i$, and we denote $J_i = (t_{i-1}, t_i)$. Fix $i \in \{1, \dots, m+1\}$; if either $\widetilde{v}(t) \geq k\ \forall t \in J_i$ or $\widetilde{v}(t) \leq -k\ \forall t \in J_i$ then $\int_{J_i} |(\varphi_k \circ v)'| dt = 0$. Otherwise, we define

$$\sigma_i = \inf\{t \in J_i; |\widetilde{v}(t)| < k\} \quad \tau_i = \sup\{t \in J_i; |\widetilde{v}(t)| < k\}.$$

By Lemma 2 we have $v \in W^{2,2}(J_i)$, then $\widetilde{v} \in C^1(J_i)$ and is uniformly continuous. By the Mean Value Theorem there exists $\xi_i \in J_i$ such that

$$\widetilde{v}'(\xi_i) = \frac{\widetilde{v}(\tau_i) - \widetilde{v}(\sigma_i)}{\tau_i - \sigma_i},$$

then by the fundamental theorem of calculus and Hölder inequality,

$$\begin{aligned}
\int_{J_i} |(\varphi_k \circ v)'| dt = \int_{\sigma_i}^{\tau_i} |(\varphi_k \circ v)'| dt &\leq \int_{\sigma_i}^{\tau_i} \left(|\widetilde{v}'(\xi_i)| + \left| \int_{\xi_i}^{t} |v''(s)| ds \right| \right) dt \\
&\leq 2k + \left(\int_a^b |v''|^2 dt \right)^{\frac{1}{2}} \int_{J_i} |t - \xi_i|^{\frac{1}{2}} dt.
\end{aligned}$$

By summing over i and taking into account that $\sum_i a_i^p \leq (\sum_i a_i)^p$ for every $p \geq 1$ and $a_i \geq 0$, we get

$$\begin{aligned}
\int_a^b |(\varphi_k \circ v)'| dt &\leq 2k(m+1) + \frac{2}{3} \left(\int_a^b |v''|^2 dt \right)^{\frac{1}{2}} \sum_{i=1}^{m+1} (t_i - t_{i-1})^{\frac{3}{2}} \\
&\leq 2k(\mathcal{H}^0(S_v \cup S_{v'}) + 1) + \frac{2}{3} \left(\int_a^b |v''|^2 dt \right)^{\frac{1}{2}} (b-a)^{\frac{3}{2}}. \qquad \square
\end{aligned}$$

Now we prove an interpolation inequality in dimension n.

THEOREM 6 *Let Q be a cube with edges of length l and parallel to the axes, $v \in GSBV^2(Q)$ and $k \in \mathbf{N}$. Then for every $i = 1, \dots, n$ the following interpolation inequality holds true*

$$\int_Q |\nabla_i (\varphi_k \circ v)| dy \leq 2k \left(\mathcal{H}^{n-1}(S_v \cup S_{\nabla v}) + l^{n-1} \right) + \frac{2}{3} l^{\frac{n}{2}+1} \left(\int_Q |\nabla^2 v|^2 dy \right)^{\frac{1}{2}}.$$

Proof By Theorems 2, 5, 3, and 4, and by the Hölder inequality, we get

$$\begin{aligned}\int_Q |\nabla_i(\varphi_k \circ v)|dy &= \int_{\pi_i(Q)} \left(\int_{Q_x} |(\varphi_k \circ v)_x{}'(t)|dt \right) dx \\ &\leq \int_{\pi_i(Q)} \left(2k(\mathcal{H}^0(S_{v_x} \cup S_{v_x{}'}) + 1) + \frac{2}{3} l^{\frac{3}{2}} \left(\int_{Q_x} |v_x{}''(t)|^2 dt \right)^{\frac{1}{2}} \right) dx \\ &\leq 2k(\mathcal{H}^{n-1}(S_v \cup S_{v'}) + l^{n-1}) + \frac{2}{3} l^{\frac{n}{2}+1} \left(\int_Q |\nabla^2 v|^2 dy \right)^{\frac{1}{2}}. \qquad \square\end{aligned}$$

5. Compactness

We state a compactness theorem for the sublevels of functional (1.3). First we consider the one-dimensional case and then the general multidimensional case.

THEOREM 7 *Assume $A \subset \mathbf{R}$ to be a bounded open set, $g \in L^2(A)$, $(z_h) \subset GSBV^2(A)$. Let $\mu, \alpha, \beta > 0$ and assume*

$$\sup_h \mathcal{F}_1(z_h, A) < +\infty.$$

Then there are a subsequence (z_{h_m}) and $z_0 \in GSBV^2(A) \cap L^2(A)$ such that, as $m \to +\infty$, $z_{h_m} \to z_0$ a.e. in A and w-$L^2(A)$ $z_{h_m} \to z_0$ strongly in $L^q(A)$ for

every $1 \leq q < 2$ $z'_{h_m} \to z'_0$ a.e. in A $z''_{h_m} \rightharpoonup z''_0$ w-$L^2(A)$.

Proof The proof can be obtained by the same argument as in [C], Theorem 2.1 and Remark 2.2, taking into account Lemma 2.

In the multidimensional case we obtain the following result.

THEOREM 8 *Let $\Omega \subset \mathbf{R}^n$ be a bounded open set, $\alpha, \beta, \mu > 0$ and let $(v_h) \subset GSBV^2(\Omega)$ be such that*

$$\sup_h \mathcal{F}(v_h) < \infty.$$

Then there are a subsequence (v_{h_m}) and $v_0 \in GSBV^2(\Omega) \cap L^2(\Omega)$ such that, as $m \to +\infty$,

$v_{h_m} \to v_0$ a.e. *in* Ω *and* w-$L^2(\Omega)$

$v_{h_m} \to v_0$ *strongly in* $L^q(\Omega)$ *for every* $1 \leq q < 2$

$\nabla v_{h_m} \to \nabla v_0$ a.e. *in* Ω

$\nabla^2 v_{h_m} \rightharpoonup \nabla^2 v_0$ w-$[L^2(\Omega)]^{n \times n}$.

Proof Since $\sup_h \|v_h\|_{L^2} < \infty$, there exist a subsequence, that we still denote by (v_h) and $v_0 \in L^2(\Omega)$ such that

$$\lim_h v_h = v_0 \qquad w - L^2(\Omega). \tag{5.1}$$

Fix $k \in \mathbf{N}$ and let $\varphi_k \in C^2(\mathbf{R})$ as in (4.1). Then we have

$$\|\varphi_k \circ v_h\|_{L^1} \leq \|v_h\|_{L^2} |\Omega|^{1/2} \qquad \forall h \in \mathbf{N}$$

and, by the interpolation Theorem 6, for every $i = 1, \ldots, n$ and for every cube $Q \subset \Omega$ with edges parallel to the axes of length l, we have

$$\sup_h \int_Q |\nabla_i(\varphi_k \circ v_h)| dy \leq c(k, l) < +\infty.$$

By the definition of $SBV(\Omega)$ and by the previous inequalities, for every $\Omega' \subset\subset \Omega$ we obtain

$$\begin{aligned} \sup_h \|\varphi_k \circ v_h\|_{BV(\Omega')} &= \sup_h \left(\int_{\Omega'} |D(\varphi_k \circ v_h)| + \int_{\Omega'} |\varphi_k \circ v_h| dy \right) \\ &\leq \sup_h \left(\int_{\Omega'} (|\nabla(\varphi_k \circ v_h)| + |\varphi_k \circ v_h|)\, dy + 2k\mathcal{H}^{n-1}(S_{v_h}) \right) < +\infty, \end{aligned} \tag{5.2}$$

hence, by the compactness theorem in $BV_{\text{loc}}(\Omega)$, by the equiboundedness in $L^\infty(\Omega)$ with respect to h of $(\varphi_k \circ v_h)$ and by a diagonalization argument, there exist a subsequence (that we do not relabel) and $w_k \in BV_{\text{loc}}(\Omega)$ such that

$$\lim_h \varphi_k \circ v_h = w_k \qquad \text{a.e. and in } L^p(\Omega) \quad 1 \leq p < +\infty, \forall k \in \mathbf{N}. \tag{5.3}$$

Set $E_k = \{|w_k| \leq k - 1\}$, then by (5.1)

$$\lim_h v_h \chi_{E_k} = v_0 \chi_{E_k} \quad w - L^2(\Omega),$$

by the properties of φ_k and by (5.3)

$$\lim_h v_h \chi_{E_k} = \lim_h (\varphi_k \circ v_h) \chi_{E_k} = w_k \chi_{E_k} \qquad \text{a.e. in } \Omega,$$

hence, since $|\varphi_k \circ v_h| \leq k$ a.e. in Ω, we get

$$w_k = v_0 \quad \text{a.e. in } E_k \ \forall k \in \mathbf{N}.$$

Since $E_k = \{|v_0| \leq k - 1\}$ and $v_0 \in L^2(\Omega)$, we have $\lim_k |\Omega \setminus E_k| = 0$ and then

$$\lim_h v_h = v_0 \qquad \text{a.e. in } \Omega, \tag{5.4}$$

hence for every i and for $\mathcal{H}^{n-1}$-a.e. $x \in \pi_i(\Omega)$ we have

$$(v_h)_x \to (v_0)_x \qquad \text{a.e. in } \Omega_x. \tag{5.5}$$

By (5.1), (5.4) and the boundedness of Ω we get $v_h \to v_0$ strongly in $L^q(\Omega)$ for $1 \leq q < 2$. By (5.2), (5.3) and (5.4), $\varphi_k \circ v_0 = w_k \in BV_{\text{loc}}(\Omega)$ $\forall k$, hence $v_0 \in GBV(\Omega)$. By using Fatou's Lemma, Fubini's Theorem, and Theorems 2, 3 and 4, for every i we get

$$\begin{aligned}
&\int_{\pi_i(\Omega)} \liminf_h \left(\int_{\Omega_x} |(v_h)_x{}''|^2 dt + \mathcal{H}^0(S_{(v_h)_x} \cup S_{(v_h)'_x}) \right) dx \\
&\leq \liminf_h \int_{\pi_i(\Omega)} \left(\int_{\Omega_x} |(v_h)_x{}''|^2 dt + \mathcal{H}^0(S_{(v_h)_x} \cup S_{(v_h)'_x}) \right) dx \\
&\leq \liminf_h \left(\int_{\Omega} |\nabla_i^2 v_h|^2 dx + \mathcal{H}^{n-1}(S_{v_h} \cup S_{\nabla_i v_h}) \right) < +\infty
\end{aligned}$$

by the assumption, so that for $\mathcal{H}^{n-1}$-a.e. $x \in \pi_i(\Omega)$ we have

$$\liminf_h \left(\int_{\Omega_x} \left(|(v_h)_x{}''|^2 + |(v_h)_x|^2 \right) dt + \mathcal{H}^0(S_{(v_h)_x} \cup S_{(v_h)'_x}) \right) < +\infty, \tag{5.6}$$

hence for $\mathcal{H}^{n-1}$-a.e. $x \in \pi_i(\Omega)$, by Theorem 7, (5.5) and the limit uniqueness, we get $(v_0)_x \in GSBV^2(\Omega_x)$. By the same argument for every $i = 1, \ldots, n$, and by Theorem 2(iv) we obtain $v_0 \in GSBV(\Omega)$ and also $\nabla v_0(x) \in \mathbf{R}^n$ for a.e. $x \in \Omega$. The property just stated allows us to use the same argument in the proof of Theorem 2.1 in [A1] for the sequence $(\nabla_i v_h)$, hence there exists $f_i \in GSBV(\Omega)$ such that, up to a subsequence,

$$\begin{aligned}
\nabla_i v_h &\to f_i && \text{a.e. in } \Omega \\
\nabla_{ij}^2 v_h &\rightharpoonup \nabla_j f_i && w - L^2(\Omega).
\end{aligned}$$

By (5.5) and the slicing argument above we know that

$$\nabla_i v_h \to \nabla_i v_0 \qquad \text{a.e. in } \Omega$$

hence $\nabla_i v_0 = f_i \in GSBV(\Omega)$ and then $v_0 \in GSBV^2(\Omega)$. □

6. Lower semicontinuity

Before stating the lower semicontinuity theorem, we prove some lemmas. The first one allows us to deal with the singular part of the mixed derivatives which cannot be estimated by simply considering any fixed one-dimensional fiber.

DEFINITION 5 *For every $(\mathcal{H}^{n-1}, n-1)$ rectifiable set $\Sigma \subset \Omega$ we denote by $\nu_\Sigma(y)$ the approximate unit normal to Σ at y, defined for $\mathcal{H}^{n-1}$ a.e. $y \in \Sigma$.*

We remark that, up to the orientation, for every $v \in GSBV^2(\Omega)$

$$\begin{aligned}
\nu_\Sigma &= \nu_v && \text{if } \Sigma \subset S_v \\
\nu_\Sigma &= \nu_{S_{\nabla v}} && \text{if } \Sigma \subset S_{\nabla v}.
\end{aligned}$$

For every fixed $\nu \in \partial B_1$ we make a conventional choice of ν_Σ by requiring $\nu_\Sigma \cdot \nu \geq 0$. Such a choice provides a unique orientation in every portion of Σ such that $\nu_\Sigma \cdot \nu \neq 0$.

LEMMA 3 *Let $v \in GSBV^2(\Omega)$. Fix $i \in \{1,\dots,n\}$. Choose $\nu,\eta \in \partial B_1$, and let $\Sigma \subset S_{\nabla_i v} \setminus S_v$ be a Borel set such that*
(i) $\nu \neq \eta, \qquad \nu \cdot \eta > \frac{3}{4},$
(ii) $\nu \cdot \nu_\Sigma(y) > \frac{3}{4}, \qquad \eta \cdot \nu_\Sigma(y) > \frac{3}{4}, \qquad \mathcal{H}^{n-1}$ a.e. $y \in \Sigma$,
(iii) $\{\nu, \eta, \mathbf{e}_i\}$ *linearly dependent.*
Then there are two disjoint Borel sets Σ_0, Σ_1 such that $\mathcal{H}^{n-1}(\Sigma \setminus (\Sigma_0 \cup \Sigma_1)) = 0$ and

$$\forall y \in \Sigma_0\ \exists x \in \pi_\nu(\Sigma_0), \quad \exists t \in S_{(v_x^\nu)'} : y = x + t\nu$$
$$\forall y \in \Sigma_1\ \exists z \in \pi_\eta(\Sigma_1), \quad \exists \tau \in S_{(v_z^\eta)'} : y = z + \tau\eta.$$

Proof Due to the assumptions (i) and (ii) the orientations induced on ν_Σ by ν and η coincide on Σ, will be explained in the following.

By assumption (ii) and by Theorem 4, if $N \subset \Sigma$ then

$$\mathcal{H}^{n-1}(N) = 0 \quad \text{iff} \quad \mathcal{H}^{n-1}(\pi_\nu(N)) = 0 \quad \text{iff} \quad \mathcal{H}^{n-1}(\pi_\eta(N)) = 0.$$

Then by Theorem 2(ii) and Theorem 4 for a.e. $y \in \Sigma$, setting $x = \pi_\nu(y)$ and $y = x + t\nu$ we have

$$t \in S_{(v_x^\nu)'} \text{ iff } (\nabla v \cdot \nu)^+(y) \neq (\nabla v \cdot \nu)^-(y),$$

hence we can choose

$$\Sigma_0 = \{y \in \Sigma; (\nabla v \cdot \nu)^+(y) \neq (\nabla v \cdot \nu)^-(y)\ \}.$$

On the other hand, if we set

$$\Sigma_1 = \{y \in \Sigma; (\nabla v \cdot \nu)^+(y) = (\nabla v \cdot \nu)^-(y)\ \},$$

since $\Sigma \subset S_{\nabla_i v} \setminus S_v$ then

$$(\nabla_i v)^+(y) = (\nabla v \cdot \mathbf{e}_i)^+(y) \neq (\nabla v \cdot \mathbf{e}_i)^-(y) = (\nabla_i v)^-(y) \qquad \mathcal{H}^{n-1} \text{ a.e. } y \in \Sigma,$$

hence by the assumptions (i) and (iii) we get

$$(\nabla v \cdot \eta)^+(y) \neq (\nabla v \cdot \eta)^-(y) \qquad \mathcal{H}^{n-1} \text{ a.e. } y \in \Sigma_1.$$

The thesis follows since, by setting $z = \pi_\eta(y), y = z + \tau\eta$, we have

$$\tau \in S_{(v_z^\eta)'} \text{ iff } (\nabla v \cdot \eta)^+(y) \neq (\nabla v \cdot \eta)^-(y). \qquad \square$$

THEOREM 9 *Let $A \subset \mathbf{R}$ be an open set, $0 < \beta \le \alpha \le 2\beta$ and $g \in L^2(A)$. Let $v_0, v_h \in GSBV^2(A)$ $(h \in \mathbf{N})$, such that*

$$\sup_{h\in\mathbf{N}} \mathcal{F}_1(v_h, A) < \infty \text{ and } v_h \to v_0 \qquad \text{a.e. in } A.$$

Then

$$\mathcal{F}_1(v_0, A) \le \liminf_h \mathcal{F}_1(v_h, A).$$

Proof Even if not stated in this form, the statement follows by [C], Theorem 3.1, taking into account that A is a union of countably many open intervals. □

LEMMA 4 *Assume $0 < \beta \le \alpha \le 2\beta$. Let $0 < \epsilon < \frac{1}{4}$, $\nu \in \partial B_1$, $i \in \{1, \ldots, n\}$, $v_0, v_h \in GSBV^2(\Omega)$ ($h \in \mathbf{N}$), $A \subset \Omega$ open set and let $\Sigma \subset A \cap (S_{\nabla_i v_0} \setminus S_{v_0})$ be a Borel set such that*

$$\nu \cdot \nu_\Sigma(y) > 1 - \frac{\epsilon}{2} \quad \mathcal{H}^{n-1}\text{-a.e. } y \in \Sigma$$

$$v_h \to v_0 \quad \text{a.e. in } \Omega$$

$$\sup_{h \in \mathbf{N}} \mathcal{F}(v_h, A) < +\infty.$$

Then

$$(1 - \epsilon)\beta \mathcal{H}^{n-1}(\Sigma) \le \liminf_h \mathcal{F}(v_h, A).$$

Proof Let $\eta \in \partial B_1$ such that $0 < |\nu - \eta| < \frac{\epsilon}{2}$ and $\{\nu, \eta, \mathbf{e}_i\}$ linearly dependent, hence

$$\eta \cdot \nu_\Sigma(y) > 1 - \epsilon \quad \mathcal{H}^{n-1}\text{-a.e. } y \in \Sigma.$$

Choose disjoint Borel sets $\Sigma_0, \Sigma_1 \subset \Sigma$ as in Lemma 3, and compact sets $K_j \subset \Sigma_j$ $(j = 0, 1)$. Let A_0, A_1 be disjoint open sets such that $K_j \subset A_j \subset A$ $(j = 0, 1)$. We can also assume that A_0 (resp. A_1) is a finite union of cubes with edges parallel or orthogonal to ν (resp. η). Then by Theorems 4 and 9, Fatou's Theorem, Fubini's Theorem and again Theorem 4 we get for any K_0, K_1 as above

$$\begin{aligned}
&\beta(1 - \epsilon)\left(\mathcal{H}^{n-1}(K_0) + \mathcal{H}^{n-1}(K_1)\right) \\
&\le \beta\left(\int_{K_0} \nu_\Sigma \cdot \nu \, d\mathcal{H}^{n-1} + \int_{K_1} \nu_\Sigma \cdot \eta \, d\mathcal{H}^{n-1}\right) \\
&= \beta\left(\int_{\pi_\nu(K_0)} \mathcal{H}^0\left((K_0)^\nu_x\right) d\mathcal{H}^{n-1}(x) + \int_{\pi_\eta(K_1)} \mathcal{H}^0\left((K_1)^\eta_z\right) d\mathcal{H}^{n-1}(z)\right) \\
&\le \int_{\pi_\nu(K_0)} \liminf_h \mathcal{F}_1\left((v_h)^\nu_x, (A_0)^\nu_x\right) d\mathcal{H}^{n-1}(x) + \\
&\qquad + \int_{\pi_\eta(K_1)} \liminf_h \mathcal{F}_1\left((v_h)^\eta_z, (A_1)^\eta_z\right) d\mathcal{H}^{n-1}(z) \\
&\le \liminf_h \left(\int_{\pi_\nu(K_0)} \mathcal{F}_1\left((v_h)^\nu_x, (A_0)^\nu_x\right) d\mathcal{H}^{n-1}(x) + \right. \\
&\qquad \left. + \int_{\pi_\eta(K_1)} \mathcal{F}_1\left((v_h)^\eta_z, (A_1)^\eta_z\right) d\mathcal{H}^{n-1}(z)\right) \\
&\le \liminf_h \left(\mathcal{F}(v_h, A_0) + \mathcal{F}(v_h, A_1)\right) \le \liminf_h \mathcal{F}(v_h, A).
\end{aligned}$$

By the arbitrariness of the compact sets K_0, K_1 and by the regularity of $\mathcal{H}^{n-1}$ the thesis follows. □

LEMMA 5 *Assume $0 < \beta \le \alpha \le 2\beta$. Let $0 < \epsilon < \frac{1}{4}$, $\nu \in \partial B_1$, $i \in \{1, \dots, n\}$, $v_0, v_h \in GSBV^2(\Omega)$ ($h \in \mathbf{N}$), $U \subset \Omega$ open set and $T \subset U \cap S_{v_0}$ Borel set such that*

$$\nu \cdot \nu_T(y) > 1 - \frac{\epsilon}{2} \quad \mathcal{H}^{n-1} - \text{a.e. } y \in T$$

$$v_h \longrightarrow v_0 \qquad \text{a.e. in } \Omega$$

$$\sup_{h \in \mathbf{N}} \mathcal{F}(v_h, U) < +\infty.$$

Then

$$(1-\epsilon)\alpha \mathcal{H}^{n-1}(T) \le \liminf_h \mathcal{F}(v_h, U).$$

Proof We can argue as in the proof of the previous lemma but in a simpler way since it is not necessary to split T into two sets due to the fact that we can use Theorem 2(iii) instead of Theorem 3(ii) and Lemma 3. □

We use the previous results to show the lower semicontinuity of $\mathcal{F}$.

THEOREM 10 *Let $\Omega \subset \mathbf{R}^n$ a bounded open set, $0 < \beta \le \alpha \le 2\beta$ and $g \in L^2(\Omega)$. Let $v_0, v_h \in GSBV^2(\Omega)$ ($h \in \mathbf{N}$), such that*

$$\sup_{h \in \mathbf{N}} \mathcal{F}(v_h) < \infty \text{ and } v_h \longrightarrow v_0 \qquad \text{a.e. in } \Omega.$$

Then

$$\mathcal{F}(v_0) \le \liminf_h \mathcal{F}(v_h).$$

Proof Fix $0 < \epsilon < \frac{1}{4}$. Let $\{\nu_k\}$ be a finite set in ∂B_1 such that for every $\nu \in \partial B_1$ there exists k such that $|\nu - \nu_k| < \frac{\epsilon}{4}$. We can find a finite family of Borel sets $\{\Sigma_j; j = 1, \dots, p+q\}$ such that

$$\Sigma_j \subset S_{v_0} \quad \text{for } j = 1, \dots, p, \quad \Sigma_j \subset S_{\nabla v_0} \quad \text{for } j = p+1, \dots, p+q,$$

$$\Sigma_j \cap \Sigma_i = \emptyset \quad \text{for } j \ne i, \quad \mathcal{H}^{n-1}((S_{v_0} \cup S_{\nabla v_0}) \setminus \cup_{j=1}^{p+q} \Sigma_j) = 0$$

$$\forall j \; \exists k_j : \; \nu_{S_{v_0} \cup S_{\nabla v_0}}(y) \cdot \nu_{k_j} > 1 - \frac{\epsilon}{2} \quad \mathcal{H}^{n-1} - \text{a.e. } y \in \Sigma_j.$$

By the regularity of $\mathcal{H}^{n-1}$ there exist disjoint compact sets $K_j \subset \Sigma_j$ such that

$$\mathcal{H}^{n-1}\big((S_{v_0} \cup S_{\nabla v_0}) \setminus \cup_{j=1}^{p+q} K_j\big) < \epsilon$$

and there exist pairwise disjoint open sets $\{A_j; j = 0, 1, \ldots, p+q\}$ such that $K_j \subset A_j$ for $j = 1, \ldots, p+q$, and

$$\int_{\Omega \setminus A_0} (|\nabla^2 v_0|^2 + \mu |v_0 - g|^2) dy < \epsilon.$$

From Lemmas 4 and 5 we get

$$\begin{aligned}
&\alpha \mathcal{H}^{n-1}(S_{v_0}) + \beta \mathcal{H}^{n-1}(S_{\nabla v_0} \setminus S_{v_0}) \\
&\leq (\alpha+\beta)\epsilon + \alpha \mathcal{H}^{n-1}(\cup_{j=1}^{p} K_j) + \beta \mathcal{H}^{n-1}(\cup_{j=p+1}^{p+q} K_j) \\
&\leq (\alpha+\beta)\epsilon + \frac{1}{1-\epsilon} \sum_{j=1}^{p+q} \liminf_h \mathcal{F}(v_h, A_j).
\end{aligned}$$

By Theorem 8 and by the lower semicontinuity of quadratic forms with respect to weak convergence in L^2

$$\int_\Omega \left(|\nabla^2 v_0|^2 + \mu |v_0 - g|^2\right) dy \leq \epsilon + \liminf_h \int_{A_0} \left(|\nabla^2 v_h|^2 + \mu |v_h - g|^2\right) dy.$$

Since $\{A_j\}$ are pairwise disjoint, summing up the previous inequalities and taking into account the arbitrariness of ϵ the thesis follows. □

REMARK By inspection of the proofs of Lemmas 4, 5 and of Theorem 10, under the same assumptions of Theorem 10 the following semicontinuity inequality can be proved

$$\begin{aligned}
&\alpha \mathcal{H}^{n-1}(S_{v_0}) + \beta \mathcal{H}^{n-1}(S_{\nabla v_0} \setminus S_{v_0}) \\
&\leq \liminf_h \left(\alpha \mathcal{H}^{n-1}(S_{v_h}) + \beta \mathcal{H}^{n-1}(S_{\nabla v_h} \setminus S_{v_h})\right).
\end{aligned}$$

Acknowledgements Work partially supported by the National Research Projects "Equazioni Differenziali e Calcolo delle Variazioni" and "Analisi Nonlineare e Calcolo delle Variazioni" of the MURST and by the CNR Project "Irregular Variational Problems" .

References

[A1] L. AMBROSIO, *Compactness for a special case of functions of bounded variation*, Boll. Un. Mat. Ital., **3-B** (7) (1989), 857–881

[A2] L. AMBROSIO, *Existence theory for a new class of variational problems*, Arch. Rational Mech. Anal., **111** (1990), 291–322.

[AFP] L. AMBROSIO, N. FUSCO & D. PALLARA: *Partial Regularity of Free Discontinuity Sets II*, preprint Dip. Matematica Pisa n.2.181.823, (1994).

[AP] L. AMBROSIO & D. PALLARA: *Partial Regularity of Free Discontinuity Sets I*, preprint Dip. Matematica Pisa n.2.180.822, (1994).

[BZ] A. BLAKE & A. ZISSERMAN, *Visual Reconstruction*, The MIT Press, Cambridge, 1987.

[B] A. BRAIDES, *Lower semicontinuity conditions for functionals on jumps and creases*, preprint, 1993.

[CLT1] M. Carriero, A. Leaci & F. Tomarelli, *Plastic free discontinuities and special bounded hessian*, C. R. Acad. Sci. Paris, **314** (1992), 595–600.

[CLT2] M. Carriero, A. Leaci & F. Tomarelli, *Special Bounded Hessian and elastic-plastic plate*, Rend. Accad. Naz. delle Scienze (dei XL), (109) **XV** (1992), 223–258.

[CLT3] M. Carriero, A. Leaci & F. Tomarelli: *Strong solution for an Elastic Plastic Plate*, Calc. Var., **2** (1994), 219-240.

[CLT4] M.Carriero, A. Leaci & F. Tomarelli: *Free gradient discontinuities*, in Proc. "Calculus of Variations, Homogeneization and Continuum Mechanics", Luminy 1993, Ed. Buttazzo, Bouchitte, Suquet, World Scientific, Singapore, 1994, 131–147.

[C] A. Coscia, *Existence result for a new variational problem in one dimensional segmentation theory*, Ann. Univ. Ferrara, **XXXVII** (1991), 185–203.

[DMS] G.Dal Maso, J.M.Morel & S. Solimini: *A variational method in image segmentation: existence and approximation results*, Acta Math., **168** (1992), 89–151.

[DS] G.David & S. Semmes: *On the singular sets of minimizers of the Mumford-Shah functional*, J. Math. Pures Appl., to appear.

[DG] E. De Giorgi, *Free discontinuity problems in calculus of variations*, Frontiers in Pure & Applied Mathematics, R. Dautray Ed., North-Holland, Amsterdam, 1991, 55–61.

[DA] E. De Giorgi & L. Ambrosio, *Un nuovo tipo di funzionale del Calcolo delle Variazioni.*, Atti Accad. Naz. Lincei, Rend. Cl. Sci. Fis. Mat. Natur., **82** (1988), 199–210.

[DCL] E. De Giorgi, M. Carriero & A. Leaci, *Existence theorem for a minimum problem with free discontinuity set*, Arch. Rational Mech. Anal., **108** (1989), 195–218.

[F] H. Federer, *Geometric Measure Theory*, Springer, Berlin, 1969.

[LS] A.Leaci & S.Solimini: *Variational problems with a free discontinuity set*, in "Geometry driven diffusion in Computer Vision", ed. B. ter Haar Romeny, Kluwer, 1994, 147–154.

[MSo] J. M. Morel & S. Solimini, *Variational Models in Image Segmentation*, Birkhäuser, Basel, 1994.

[MSh] D. Mumford & J. Shah, *Optimal approximation by piecewise smooth functions and associated variational problems*, Comm. Pure Appl. Math., **XLII** (1989), 577–685.

[R] B. ter Haar Romeny ed.: *Geometry driven diffusion in Computer Vision*, Kluwer, Dordrecht, 1994.

[S] G. Savaré, *On the positive part of $W^{s,p}(\Omega)$ functions for $s > 1$*, Pubbl. I.A.N. - C.N.R., Pavia, 1994.

[T] F. Tomarelli, *Special Bounded Hessian and partial regularity of equilibrium for a plastic plate*, in "Developments in PDE and Applications to Mathematical Physics", Eds. G. Buttazzo, G. P. Galdi, L. Zanghirati, Plenum Press, N. Y., 1992, 235–240.

[Z] W. P. Ziemer, *Weakly Differentiable Functions*, Springer, N.Y., 1988.

Progress in Nonlinear Differential Equations
and Their Applications, Vol. 25

Optimal Approximation by Piecewise Constant Functions

ITALO TAMANINI

Dipartimento di Matematica, Università di Trento,
38050 Povo, Trento, ITALY

1. Introduction

In their seminal 1989 paper [8], D. Mumford and J. Shah proposed a variational approach to image segmentation in Computer Vision Theory, and studied in particular the following problem:

Given an open rectangle $R \subset \mathbb{R}^2$, a function g continuous on the closure $\bar{R}$ of R, and a positive coefficient ν, find a finite set $\Gamma = \{\gamma_1, \ldots, \gamma_n\}$ of C^2 arcs contained in $\bar{R}$, meeting each other only at their end-points, and minimizing the following functional

$$E(\Gamma) = \sum_{i=1}^{N} \iint_{R_i} |a_i - g(x,y)|^2 dxdy + \nu \cdot \text{length}(\Gamma)$$

where $R_1, \ldots, R_N$ denote the connected components of $R \backslash \Gamma$, a_i is the average of g on R_i, i.e. $a_i = ⨏\!\!⨏_{R_i} g(x,y)dxdy$, and length$(\Gamma)$ is the sum of the lengths of the arcs γ_j.

In the usual interpretation of this problem in the context of image segmentation, g represents the "grey-level" of a picture (assigning a value, say in $[0,1]$, to each point in R: e.g., $g = 0$ corresponds to "white", while $g = 1$ correspond to "black"); we are required to mark sharply (by means of the arcs γ_j) the significant discontinuities of g (corresponding to the "edges" or "contours" of the objects in the picture), and to approximate g by a uniform level a_i in each element of the induced decomposition $\{R_i\}$. The optimality criterion leading to the above minimization problem is that we require a good approximation without creating too many discontinuities. It is shown in [8] that E is the restriction to piecewise constant functions of a more general functional, which allows for non-uniform approximation inside each region R_i: see also [2], [3] and [7].

Chapter 5 of [8] contains the proof of the existence of a minimizing configuration of curves, including classification of singular points and estimates of the cardinality of the associated decomposition. The proof is in the spirit of Geometric Measure Theory: for fixed N (number of elements of a decomposition), they prove that $E = E_N$ admits an optimal configuration; on the other hand, they show that for N large enough (depending on the data) the least value of E_N cannot be improved by increasing N itself. This last assertion follows by a typical 2-dimensional argument, based on the use of Gauss-Bonnet theorem.

A different, more elementary proof of the same result was independently obtained by J. M. Morel and S. Solimini: they also employ very special properties of 2-dimensional sets. See e.g. [7], Chapter 5.

In higher dimensions new arguments seem to be needed: we first prove the existence of "optimal partitions" of a domain $\Omega \subset \mathbb{R}^n$ by using a powerful compactness theorem. Simple "split and merge" techniques, coupled with the isoperimetric inequality, then yield basic estimates and imply in particular a first "topological regularization" of the partitioning sets. Further properties (regularity, finiteness) can be deduced by "blow-up" methods. We will try to explain these methods, state some sample results about the n-dimensional version of the preceding problem, and discuss the relevance of the (integrability) assumptions on the datum g to the expected results.

2. Multidimensional segmentation

In the following we will consider the minimization problem

$$F(C,u) = \lambda \int_{\Omega\setminus C} |u-g|^p dx + \mathcal{H}^{n-1}(C \cap \Omega) \to \min \tag{1}$$

where Ω (an open subset of $\mathbb{R}^n$), $\lambda > 0$, $p \geq 1$ and $g \in L^p(\Omega)$ are given, and where any "admissible pair" (C, u) consists of a closed subset C of $\mathbb{R}^n$ and a C^1 function u with vanishing gradient everywhere in $\Omega\setminus C$. Roughly speaking, we are seeking an "optimal" approximation of g by a function u which is constant in each (open, connected) component of $\Omega\setminus C$, and discontinuous across C — the optimality criterion being precisely the minimization of the functional F (a kind of "compromise" between good approximation and small discontinuity set, with respect to the Hausdorff $(n-1)$-dimensional measure $\mathcal{H}^{n-1}$).

We restrict ourselves to the case $n \geq 2$, simply because in one real variable our results — related to the segmentation of 1-D signals, quite important in the applications — can be deduced from simpler, direct arguments. Notice that the true variable in (1) is C: indeed, given C, the optimal $u = u_C$ is easily found to be

$$u_{C|_A} = t_A \quad \text{s.t.} \quad \int_A |t_A - g(x)|^p dx = \min_{t\in\mathbb{R}} \int_A |t-g(x)|^p dx$$

for all connected components A of $\Omega\setminus C$.

We want to apply the direct method of the Calculus of Variations to the minimum problem (1); however, due to the difficulty of proving the lower semicontinuity of $\mathcal{H}^{n-1}$ (with respect to, say, the Hausdorff metric on closed sets) on minimizing sequences, we first shift to a weaker, more tractable formulation. To this aim, we extend the (rather strong) notion of "decomposition of Ω" related to the above formulation: in fact, if $F(C,u) < \infty$, then C decomposes Ω in a countable family of open, connected, mutually disjoint regions, covering almost all of Ω and separated by "interfaces" of finite $(n-1)$-dimensional measure. Instead, we say that $\mathcal{U}$ is a Caccioppoli partition of Ω (briefly: $\mathcal{U} \in CP(\Omega)$) iff there is a sequence $\{U_i\}$ of measurable subset of $\mathbb{R}^n$ s.t.

$$(i) \qquad \mathcal{U} = \{U_i : i \in \mathbb{N}\}$$

$$\begin{array}{ll} (ii) & U_i = U_i(1) \quad \forall i \in \mathbb{N} \\ (iii) & U_i \cap U_j = \emptyset \quad \text{if } i \neq j \\ (iv) & |\Omega \backslash \bigcup_{i=1}^{\infty} U_i| = 0 \\ (v) & \sum_{i=1}^{\infty} P(U_i, A) < \infty \quad \forall A \text{ open } \subset\subset \Omega \end{array} \tag{2}$$

Here, $|U|$ denotes the Lebesgue measure of $U \subset \mathbb{R}^n$, $U(\alpha)$ is the set of points of density $\alpha \in [0,1]$ for U, and $P(U, A)$ is the perimeter of U in A, i.e. (see e.g. [4]):

$$U(\alpha) = \left\{ x \in \mathbb{R}^n : \lim_{r \to 0} |U \cap B_{x,r}| \, / \, |B_{x,r}| = \alpha \right\} \tag{3}$$

$$P(U, A) = \sup \left\{ \int_U \operatorname{div}\Phi(x) dx : \Phi \in C_0^1(A; \mathbb{R}^n), |\Phi(x)| \leq 1 \quad \forall x \in A \right\} \tag{4}$$

Of course, $B_{x,r}$ is the euclidean n-ball of centre x and radius $r > 0$; when $x = 0$ we simply write B_r instead of $B_{0,r}$ and put $\omega_n = |B_1|$. We call any sequence $\{U_i\}$ as above an "arrangement" of $\mathcal{U}$, and put $P(\mathcal{U}, A) = \frac{1}{2} \sum_{i=1}^{\infty} P(U_i, A)$, the "total perimeter" of $\mathcal{U}$ in A.

Assumption (ii) above is made essentially to have a definite representation of the partitioning sets — which otherwise would be defined only almost-everywhere. We will see later on that, thanks to (ii), the elements of an optimal partition are open sets. Anyway, we remark that if the sequence $\{U_i\}$ satisfies (iii) (or even $|U_i \cap U_j| = 0 \quad \text{if } i \neq j$), (iv) and (v), then by just replacing each U_i by $U_i(1)$ we are led to an equivalent Caccioppoli partition. Notice that instead of (ii) we could merely require that $U_i = U_i(1) \cap \Omega$, in which case the U_i's are subsets of Ω. In addition, we can certainly assume that $\emptyset \in \mathcal{U}$: this is formally necessary when $\mathcal{U}$ is finite, in which case any arrangement $\{U_i\}$ of $\mathcal{U}$ will contain only a finite number of terms different from the empty set. We will occasionally use these facts in the sequel, without mentioning them explicitly.

We say that:

- $\mathcal{U}$ is locally finite in Ω iff to each $x \in \Omega$ corresponds a suitable radius $r > 0$ and a positive integer m s.t.

$$U_i \cap B_{x,r} = \emptyset \text{ if } i > m$$

- $\mathcal{U}$ coincides with $\mathcal{V}$ in A, A open $\subset \Omega$ (written $\mathcal{U} = \mathcal{V}$ in A) iff arrangements $\{U_i\}$ and resp. $\{V_i\}$ can be found s.t. $U_i \cap A = V_i \cap A \quad \forall i \in \mathbb{N}$.

- $\mathcal{U}$ is a compact variation of $\mathcal{V}$ in A iff there exists a compact subset D of A s.t. $\mathcal{U} = \mathcal{V}$ in $\Omega \backslash D$.

- $\mathcal{U}_h$ converges to $\mathcal{U}$ locally in Ω (written $\mathcal{U}_h \to \mathcal{U}$ in $L^1_{\text{loc}}(\Omega)$) iff arrangements $\{U_{h,i}\}$ and resp. $\{U_i\}$ can be found s.t.

$$\lim_{h \to \infty} \int_A \left| \chi_{U_{h,i}} - \chi_{U_i} \right| dx = 0 \quad \forall i \in \mathbb{N} \text{ and } \forall A \text{ open } \subset\subset \Omega.$$

Here, χ_U is the characteristic function of U. We now state the main, general results about Caccioppoli partitions; for the proof see [1].

THEOREM 1 *(Compactness).*
Let $\mathcal{U}_h \in CP(\Omega)$ *be s.t.* $\sup_{h\in\mathbb{N}} P(\mathcal{U}_h, A) < \infty$ $\forall A$ *open* $\subset\subset \Omega$. *Then there is* $\mathcal{U} \in CP(\Omega)$ *and a subsequence* $\{\mathcal{U}_{k(h)}\}$ *of* $\{\mathcal{U}_h\}$ *s.t.* $\mathcal{U}_{k(h)} \to \mathcal{U}$ *in* $L^1_{\text{loc}}(\Omega)$.

THEOREM 2 *(Semicontinuity). If* $\mathcal{U}_h \to \mathcal{U}$ *in* $L^1_{\text{loc}}(\Omega)$, *with* $\mathcal{U}_h, \mathcal{U} \in CP(\Omega)$, *then* $P(\mathcal{U}, A) \le \liminf_{h\to\infty} P(\mathcal{U}_h, A)$ $\forall A$ *open* $\subset \Omega$.

THEOREM 3 *(Structure). If* $\{U_i\}$ *is any arrangement of* $\mathcal{U} \in CP(\Omega)$, *then*

$$\Omega = \left[\bigcup_{i=1}^{\infty}(\Omega \cap U_i)\right] \cup \left[\bigcup_{i<j}(\Omega \cap U_i(1/2) \cap U_j(1/2))\right] \cup N$$

with $N \subset \Omega$, $\mathcal{H}^{n-1}(N) = 0$.

The natural extension of the segmentation functional (1) can now be defined:

$$G(\mathcal{U}, u) = P(\mathcal{U}, \Omega) + \lambda \int_\Omega |u - g|^p\, dx \tag{5}$$

Here, the pair $(\mathcal{U}, u)$ is a "weighted Caccioppoli partition" of Ω (briefly: $(\mathcal{U}, u) \in WCP(\Omega)$), i.e. $\mathcal{U} \in CP(\Omega)$ and u is a real function, defined a.e. in Ω, constant on each element U of $\mathcal{U}$. Sometimes we denote this constant value by $u(U)$. We point out that $u(U)$ can coincide with $u(V)$ even when $U, V \in \mathcal{U}$ have a "common interface" inside Ω, i.e. when $\mathcal{H}^{n-1}(U(1/2) \cap V(1/2) \cap \Omega) > 0$. Of course, $(\mathcal{U}, u)$ will coincide with $(\mathcal{V}, v)$ in the open set $A \subset \Omega$ iff $\mathcal{U} = \mathcal{V}$ in A and $u(x) = v(x)$ for almost all $x \in A$. Similarly, $(\mathcal{U}_h, u_h) \to (\mathcal{U}, u)$ locally in A means that $\mathcal{U}_h \to \mathcal{U}$ in $L^1_{\text{loc}}(A)$ and $u_h(x) \to u(x)$ for a.a. $x \in A$. With respect to this convergence, the previous Compactness Theorem 1 extends to weighted Caccioppoli partitions as well. See [11], Theorem 3.1.

As a straightforward consequence of the aforementioned results we have the following

PROPOSITION 1 *(Existence). In the usual assumptions (i.e. if* Ω *is open in* $\mathbb{R}^n$, $n \ge 2$, $\lambda > 0$, $p \ge 1$, *and* $g \in L^p(\Omega)$*), there is* $(\mathcal{W}, w) \in WCP(\Omega)$ *s.t.*

$$G(\mathcal{W}, w) \le G(\mathcal{U}, u) \quad \forall (\mathcal{U}, u) \in WCP(\Omega) \tag{6}$$

We call such a $\mathcal{W}$ *an "optimal partition" of* Ω.

Indeed, notice that $0 \le \inf\{G(\mathcal{U}, u) : (\mathcal{U}, u) \in WCP(\Omega)\} \le G(\{\mathbb{R}^n, \emptyset\}, 0) = \lambda \int_\Omega |g|^p\, dx < \infty$. By compactness, we can extract from any minimizing sequence a convergent subsequence, whose limit (by semicontinuity and Fatou's Lemma) is the required solution. Therefore, the simple integrability condition $g \in L^p(\Omega)$

entails the existence of an absolute minimizer $(\mathcal{W}, w)$ of G. The following upper bound on the total perimeter of $\mathcal{W}$ inside a ball $B = B_{x,r} \subset \Omega$ follows easily from a simple "split & merge" technique:

$$P(\mathcal{W}, B) \leq \mathcal{H}^{n-1}(\partial B) + \lambda \int_B |g|^p \, dx \tag{7}$$

For, given an arrangement $\{W_i\}$ of $\mathcal{W}$ with $W_1 = \emptyset$, we define $U_1 = B$, $u(U_1) = 0$; $U_i = W_i \backslash \bar{B}$, $u(U_i) = w(W_i) \quad \forall i \geq 2$. By comparison, see (6), we get immediately (7). If in addition $g \in L^{np}_{\text{loc}}(\Omega)$, we get from Hölder's inequality

$$P(\mathcal{W}, B_{x,r}) \leq \omega_n \left(n + \lambda \omega_n^{-1/n} \|g\|^p_{L^{np}(B_{x,r})} \right) r^{n-1},$$

hence for all $x \in \Omega$:

$$\limsup_{r \to 0} r^{1-n} P(\mathcal{W}, B_{x,r}) \leq n\omega_n. \tag{8}$$

If we use the spherical surface $\partial B_{x,r}$ to split only one element $W \in \mathcal{W}$, we obtain by a similar argument

$$|w(W)| \, |W \cap B|^{1/p} \leq (\lambda^{-1} \mathcal{H}^{n-1}(W \cap \partial B))^{1/p} + 2\|g\|_{L^p(W \cap B)} \tag{9}$$

where again $B = B_{x,r} \subset \Omega$ (see [11], Lemma 5.1). Clearly, (7) and (9) hold if $(\mathcal{W}, w)$, instead of being an absolute minimizer of G satisfying (6), is merely a minimizer of G_g in $B_{x,R}$ for some $R \in (r, \text{dist}(x, \partial\Omega))$, according to the following definition:

$$\begin{aligned} &(\mathcal{W}, w) \in WCP(\Omega) \quad \textit{is a minimizer of } G_g \textit{ in } B = B_{x,R} \subset \Omega \textit{ iff} \\ &G_g(\mathcal{W}, w; B) < \infty \quad \textit{and } G_g(\mathcal{W}, w; B) \leq G_g(\mathcal{U}, u; B) \end{aligned} \tag{10}$$

for all compact variations $(\mathcal{U}, u)$ of $(\mathcal{W}, w)$ in $B_{x,R}$, i.e. coinciding with $(\mathcal{W}, w)$ outside $\bar{B}_{x,t}$ for some $t \in (0, R)$,

where

$$G_g(\mathcal{U}, u; B) = P(\mathcal{U}, B) + \lambda \int_B |u - g|^p \, dx \tag{11}$$

Suppose now that $\Omega = \mathbb{R}^n$, $g = 0$ and let $(\mathcal{W}, w)$ be a minimizer of G_0 in B_R, for all $R > 0$. Fix $W \in \mathcal{W}$ and $R > 0$ s.t. $W \cap B_R \neq \emptyset$ and define $\alpha(r) = |W \cap B_r|$, so that α is non-decreasing, with $0 < \alpha(r) \leq \omega_n r^n$ and derivative $\alpha'(r) = \mathcal{H}^{n-1}(W \cap \partial B_r)$ for a.a. $r > R$. From (9) we get for a.a. $r > R$:

$$\lambda \, |w(W)|^p \, \alpha(r) \leq \alpha'(r)$$

which, according to the stated assumptions, gives immediately $w(W) = 0$. Thus $w = 0$ a.e. in $\mathbb{R}^n$. Furthermore, if $\mathcal{U} \in CP(\mathbb{R}^n)$ is a compact variation of $\mathcal{W}$ in B_R, then from (10), (11) (with $u = w = 0$) we get

$$P(\mathcal{W}, B_R) < \infty \text{ and } P(\mathcal{W}, B_R) \leq P(\mathcal{U}, B_R) \text{ for all } R > 0 \tag{12}$$

i.e. $\mathcal{W}$ is (by definition) a "minimal partition" of $\mathbb{R}^n$. As we will see later on, this fact is of basic importance in the study of the local finiteness of optimal partitions.

A more refined argument, still based on SPLIT & MERGE, gives us a basic Clearing-out Lemma: we outline the argument in the proof of the following

PROPOSITION 2 *If $(\mathcal{W}, w)$ is a minimizer of G (as in (6)), with $g \in L^p(\Omega) \cap L^{np}_{\mathrm{loc}}(\Omega)$, then each $W \in \mathcal{W}$ is open in Ω.*

Indeed, fix $W \in \mathcal{W}$ with $W \cap \Omega \neq \emptyset$, fix an arrangement $\{W_i\}$ of $\mathcal{W}$ with $W_1 = W$, and put $t = w(W)$, $V = \bigcup_{i=2}^{\infty} W_i$. Fix a point of $W \cap \Omega$, and assume for simplicity it is the origin of $\mathbb{R}^n$. We can then find $R > 0$ s.t. $B_R \subset \Omega$ and

$$2\lambda \left(\int_{B_R} |t - g(x)|^{np} \, dx \right)^{1/n} \leq n\omega \qquad (\text{since } g \in L^{np}_{\mathrm{loc}}(\Omega)) \tag{13}$$

$$|V \cap B_R| = |B_R \backslash W| \leq \omega_n 8^{-n} R^n \qquad (\text{since } W = W_1(1)) \tag{14}$$

Recall (ii) of (2). We claim that $|V \cap B_{R/2}| = 0$: from this it follows at once that $B_{R/2} \subset W$, thus proving that $W \cap \Omega$ is open. To prove the claim, define

$$\alpha(r) = |V \cap B_r| = \int_0^r dt \int_{\partial B_t} \chi_V \, d\mathcal{H}^{n-1}$$

and assume that $\alpha(r) > 0$ in the interval $(R/2, R)$ (otherwise there is nothing to prove). For fixed $r \in (R/2, R)$ define $U_1 = W_1 \cup B_r$, $U_i = W_i \backslash \bar{B}_r \quad (i \geq 2)$; $u(U_i) = w(W_i) \quad (i \in \mathbb{N})$. This gives a compact variation $(\mathcal{U}, u)$ of $(\mathcal{W}, w)$ in B_R, and we deduce from (6) that for a.a. $r \in (R/2, R)$:

$$\begin{aligned}
0 &\leq P(\mathcal{U}, B_R) - P(\mathcal{W}, B_r) + \lambda \int_{B_R} (|u - g|^p - |w - g|^p) \, dx \\
&\leq \int_{\partial B_r} \chi_V \, d\mathcal{H}^{n-1} - P(V, B_r) + \lambda \int_{V \cap B_r} |t - g(x)|^p \, dx \\
&\leq 2 \int_{\partial B_r} \chi_V \, d\mathcal{H}^{n-1} - P(V \cap B_r, \mathbb{R}^n) + \lambda \left(\int_{B_R} |t - g(x)|^{np} \, dx \right)^{1/n} \cdot \\
&\quad \cdot |V \cap B_r|^{(n-1)/n} \\
&\leq 2 \int_{\partial B_r} \chi_V \, d\mathcal{H}^{n-1} - n\omega^{1/n} |V \cap B_r|^{(n-1)/n} + \frac{1}{2} n\omega_n |V \cap B_r|^{(n-1)/n} \\
&\leq 2\alpha'(r) - \frac{1}{2} n\omega_n^{1/n} |V \cap B_r|^{(n-1)/n}
\end{aligned}$$

where the last but one inequality follows from the isoperimetric inequality and (13) above. In other words, for a.a. $r \in (R/2, R)$ we have

$$4^{-1} \omega_n^{1/n} \leq n^{-1} \alpha(r)^{(1-n)/n} \alpha'(r) = (\alpha(r)^{1/n})'$$

which can be integrated on $[R/2, R]$ to give, on the account of (14): $\alpha(R/2) = 0$, thus proving our claim.

In conclusion, each element W of the optimal partition $\mathcal{W}$ is open in Ω, and it is now a simple matter to check that (K, w) with $K = \overline{\Omega \backslash \bigcup \mathcal{W}}$ is a minimizer of F in (1), i.e.

$$F(K, w) \leq F(C, u) \text{ for all admissible pairs } (C, u) \tag{15}$$

Moreover

$$\mathcal{H}^{n-1}(K\cap\Omega)=P(\mathcal{W},\Omega) \text{ i.e. } F(K,w)=G(\mathcal{W},w) \tag{16}$$

(this uses the Structure Theorem 3, see [11], Prop. 5.4).

To sum up: we have obtained an optimal segmentation K (i.e. a minimizer of the functional F defined on "regular" decompositions of Ω) by

a) proving the existence of an optimal partition $\mathcal{W}$

b) showing that its elements are open in Ω.

Notice that a) was derived under the sole hypothesis $g\in L^p(\Omega)$, while for b) we assumed in addition that $g\in L^{np}_{\mathrm{loc}}(\Omega)$. Actually, there are functions $g\in L^q(\Omega)$ for all $q<np$, for which:

c) not all elements of an optimal partition $\mathcal{W}$ are open

d) no minimizer of F can exist.

Simple examples are constructed in [1] and [11].

Having thus solved the existence problem, we go on with the study of the minimizers of F and G. First we remark the following fact, implicit in the proof of Prop. 2: if an arbitrary element W of an optimal partition $\mathcal{W}$ occupies a large part of a ball $B_R\subset\Omega$ (see (14)), then it fills completely the concentric ball of radius $R/2$! By a closer inspection of the proof, we can show the following

LEMMA 1 *(Clearing-out). Let $(\mathcal{W},w)$ be a minimizer of G_g in $B_R\subset\Omega$, with $g\in L^{np}_{\mathrm{loc}}(\Omega)$ (recall (10), (11)). Fix an arrangement $\{W_i\}$ of $\mathcal{W}$ and put $t_i=w(W_i)$. For $m\in\mathbb{N}$, define*

$$\eta=\eta(n,m)=\omega_n 4^{-n}(m+1)^{-n} \tag{17}$$

and choose $r\in(0,R)$ s.t.

$$2m\lambda\cdot\max_{1\le i\le m}\left(\int_{B_r}|t_i-g(x)|^{np}\,dx\right)^{1/n}\le n\omega_n^{1/n}. \tag{18}$$

If $|B_r\backslash\bigcup_{i=1}^m W_i|\le\eta r^n$, then $\left|B_{r/2}\backslash\bigcup_{i=1}^m W_i\right|=0$.

See [11], lemma 5.3.

There are several useful consequences of this lemma. The first one concerns the local finiteness of optimal partitions. We first study this question for minimal partitions, recall (12). Thus, let $\{V_i\}$ be an arrangement of $\mathcal{V}\in CP(\Omega)$, and assume that in a certain ball $B_R\subset\Omega$ there holds:

$$P(\mathcal{V},B_R)<\infty \text{ and } P(\mathcal{V},B_R)\le P(\mathcal{U},B_R) \tag{19}$$

for all compact variations $\mathcal{U}$ of $\mathcal{V}$ in B_R. By blowing-up $\mathcal{V}$ near 0 we get a conical minimal partition $\mathcal{C}$ of $\mathbb{R}^n$: indeed, by defining $\mathcal{V}_h=\{V_{h,i}\}$ with $V_{h,i}=\{hx:x\in V_i\}$ $(h,i\in\mathbb{N})$ we check immediately that $\mathcal{V}_h$ is a minimal partition of B_{hR}. As in (7), we have a locally uniform upper bound on the perimeters:

$$P(\mathcal{V}_h,B_t)=h^{n-1}P(\mathcal{V},B_{h^{-1}t})\le h^{n-1}\mathcal{H}^{n-1}(\partial B_{h^{-1}t})=n\omega_n t^{n-1} \tag{20}$$

$\forall t > 0$, $\forall h > t/R$. By compactness, a subsequence of $\mathcal{V}_h$ converges to $\mathcal{C}$, and it is easily seen that $\mathcal{C}$ is minimal in $\mathbb{R}^n$. A more involved argument based on a monotonicity formula (see [5], theor. 7) shows that $\mathcal{C} = \{C_i\}$, where any C_i is a cone of vertex 0 (i.e., χ_{C_i} is positively homogeneous of degree 0). It turns out that $\mathcal{C}$ is finite. We prove this by induction on the dimension n.

For $n = 2$, each element C_i is an open cone (same argument of Prop. 2) and the interfaces $\partial C_i \cap \partial C_j$ are half-lines issuing from 0. As in (7) we have

$$P(\mathcal{C}, B_1) = \frac{1}{2}\sum_i P(C_i, B_1) \leq \mathcal{H}^1(\partial B_1) = 2\pi$$

hence no more than 6 distinct C_i are non-empty. Indeed, it is well-known that the number of non-empty elements of a conical minimal partition of the plane cannot be greater than 3 (elementary trigonometry implies that the angle between any two consecutive boundary half-lines has to measure 120° at least).

Now suppose that the result is true in $\mathbb{R}^n$, for some $n \geq 2$, and let $\mathcal{C} = \{C_i\}$ be a conical minimal partition of $\mathbb{R}^{n+1}$. Fix a point $x_0 \in \mathbb{R}^{n+1}$ with $|x_0| = 1$. By blowing-up near x_0 we find a cylindrical minimal partition $\mathcal{Q}$ of $\mathbb{R}^{n+1}$ (see [5], Theor. 8), i.e. (up to a rotation) $\mathcal{Q} = \{Q_i\}$ with $Q_i = C'_i \times \mathbb{R}$, C'_i a cone $\subset \mathbb{R}^n$ $\forall i$. $\mathcal{C}' = \{C'_i\}$ is then a conical, minimal partition of $\mathbb{R}^n$, hence finite by our inductive hypothesis, hence $\mathcal{Q}$ is finite, too, i.e. there exists $m \in \mathbb{N}$ s.t. $Q_i = \emptyset$ for $i > m$. Now, by construction, $\mathcal{Q}$ is the limit of a subsequence (not relabeled) of the expanded sequence $\mathcal{C}_h = \{C_{h,i}\}$ with $C_{h,i} = \{x_0 + h(x - x_0) : x \in C_i\}$, hence

$$\lim_{h\to\infty} |B_{x_0,1} \backslash (C_{h,1} \cup \cdots \cup C_{h,m})| = 0$$

i.e. $|B_{x_0,1} \backslash (C_{h,1} \cup \cdots \cup C_{h,m})| < \eta(n+1, m)$ if h is large enough, see (17). From the Clearing-out Lemma 1 (with $g = 0$, see (10)–12)) we get for the same h:

$$h^n \left|B_{x_0,2^{-1}h^{-1}} \backslash (C_1 \cup \cdots \cup C_m)\right| = \left|B_{x_0,2^{-1}} \backslash (C_{h,1} \cup \cdots \cup C_{h,m})\right| = 0$$

thus showing that $\mathcal{C}$ is finite — being finite in a neighborhood of each point of the unit sphere of $\mathbb{R}^{n+1}$. In conclusion:

PROPOSITION 3 *If $\mathcal{C}$ is a conical minimal partition of $\mathbb{R}^n$, then $\mathcal{C}$ is finite.*

Going back to the partition $\mathcal{V}$ satisfying (19) and recalling that the enlarged partition $\mathcal{V}_h$ converges to a partition $\mathcal{C}$ of this type, we obtain by a new application of Clearing-out Lemma 1 a radius $r \in (0, R)$ s.t. $\mathcal{V}$ is finite in B_r. Actually, what is true is that $\mathcal{V}$ and $\mathcal{C}$ have the same cardinality in B_r.

Now we come to the general case of a minimizer $(\mathcal{W}, w)$ of G, as in Prop. 1, with $g \in L^p(\Omega) \cap L^{np}_{\text{loc}}(\Omega)$, and show that $\mathcal{W}$ is locally finite in Ω. If B_R is any ball $\subset\subset \Omega$ (we assume for simplicity that its center is located at the origin), then clearly $(\mathcal{W}, w)$ is a minimizer of G_g in B_R (see (10), (11)), and $g \in L^{np}(B_R)$. We dilate sets in the usual way

$$U_h = \{hx : x \in U\}$$

and rescale functions as follows

$$u_h(x) = h^{-1/p}u(h^{-1}x)$$

With this notation, if $\mathcal{W}_h = \{W_h : W \in \mathcal{W}\}$, then for all $h \in \mathbb{N}$, $(\mathcal{W}_h, w_h)$ is a minimizer of G_{g_h} in $B_{hR} \subset\subset \Omega_h$. In addition, $g_h \to 0$ in $L^p_{\rm loc}(\mathbb{R}^n)$. By compactness, we find a subsequence converging to $(\mathcal{W}_\infty, w_\infty) \in WCP(\mathbb{R}^n)$, and it follows from the next proposition that $(\mathcal{W}_\infty, w_\infty)$ is a minimizer of G_0 (i.e. of G_{g_∞} with $g_\infty = 0$) in B_r for all $r > 0$. By the argument immediately following (11) we deduce that $w_\infty = 0$ and $\mathcal{W}_\infty$ is minimal in $\mathbb{R}^n$, see (12), hence finite in a neighborhood of 0. By the usual application of the Clearing-out Lemma, we deduce that $\mathcal{W}$ is finite in a neighborhood of 0, too.

The minimality of $(\mathcal{W}_\infty, w_\infty)$ quoted above is a consequence of the following

PROPOSITION 4 *(Convergence). Let $(\mathcal{W}_h, w_h)$ be a minimizer of G_{g_h} in $B = B_{x,R} \subset\subset \Omega$, with $g_h \in L^p(\Omega)$ $\forall h \in \mathbb{N}$. If*

$$\begin{aligned} \mathcal{W}_h \to \mathcal{W}_\infty &\quad in\ L^1_{\rm loc}(\Omega) \\ w_h \to w_\infty &\quad a.e.\ in\ \Omega \\ g_h \to g_\infty &\quad in\ L^p_{\rm loc}(\Omega) \end{aligned}$$

with $(\mathcal{W}_\infty, w_\infty) \in WCP(\Omega)$, then $(\mathcal{W}_\infty, w_\infty)$ is a minimizer of G_{g_∞} in B.

Its proof can be found in [11], Theor. 6.1. In conclusion, we can state the following

PROPOSITION 5 *If $(\mathcal{W}, w)$ is a minimizer of G (see (6)), with $g \in L^p(\Omega) \cap L^{np}_{\rm loc}(\Omega)$, then $\mathcal{W}$ is locally finite in Ω.*

Again, the hypotheses are sharp: an example with $g \in L^p(\Omega)$ $\forall q < np$ is shown in [1].

We now interpret the preceding results in terms of optimal segmentations, i.e. solutions of $F \to$ min. We saw before (see Prop. 1, (15) and (16)) that if $(\mathcal{W}, w)$ is a minimizer of G with $g \in L^p(\Omega) \cap L^{np}_{\rm loc}(\Omega)$ and if $K = \overline{\Omega \setminus \bigcup \mathcal{W}}$, then (K, w) is a minimizer of F; conversely, given any such (K, w) and defining

$$\mathcal{W} = \{V(1) : V \text{ is a connected component of } \Omega \setminus K\}, \tag{21}$$

we get a minimizer $(\mathcal{W}, w)$ of G, which in turn gives rise to a new, "more regular" (see (24), (26) below) minimizer (K_1, w_1) of F: we have $K_1 \subset K$, $K_1 = \overline{K_1 \cap \Omega}$, $\mathcal{H}^{n-1}((K \setminus K_1) \cap \Omega) = 0$, $w_1 = w$ in $\Omega \setminus K$. See [11], Prop. 5.4. As a corollary of Prop. 5 we then have that if (K, w) is a minimizer of F with $g \in L^p(\Omega) \cap L^{np}_{\rm loc}(\Omega)$, then the family of connected components of $\Omega \setminus K$ is locally finite in Ω. In particular, $w \in L^\infty_{\rm loc}(\Omega)$.

An interesting consequence of the finiteness result is the following lower volume density estimate: in the preceding assumptions, if $x \in \partial V \cap \Omega$ with V a connected component of $\Omega \setminus K$, then

$$\liminf_{r \to 0} \omega_n^{-1} r^{-n} |V \cap B_{x,r}| > 0 \tag{22}$$

This is because for a suitable $R > 0$, $B_{x,R} \setminus K$ has a finite number (say, $m+1$) of connected components, one of which is $V \cap B_{x,R}$, e.g.:

$$B_{x,R} \setminus K = B_{x,R} \cap (V_1 \cup \ldots \cup V_{m+1})$$

with V_i open connected $\forall i$ and $V_{m+1} = V$. Were (22) false, we would find arbitrarily small $r > 0$ for which $|B_{x,r} \setminus (V_1 \cup \ldots \cup V_m)| < \eta(n,m) r^n$ and the Clearing-out Lemma would imply that V has measure zero in a neighborhood of x, contrary to the fact that V is open and $x \in \partial V$.
The "symmetric estimate"

$$\limsup_{r \to 0} \omega_n^{-1} r^{-n} |V \cap B_{x,r}| < 1 \tag{23}$$

cannot hold for all minimizers of F: this is because if (K,w) minimizes F, N is closed $\subset \mathbb{R}^n$ with $\mathcal{H}^{n-1}(N \cap \Omega) = 0$, and $v = w_{|\Omega \setminus (K \cup N)}$, then $(K \cup N, v)$ minimizes F, too.
Nevertheless, (23) holds true for the "essential minimizers" (K_1, w_1) considered above, see formula (21) and the remark following it.
Indeed, for any optimal segmentation K which arises from an optimal partition $\mathcal{W}$ as explained above (i.e. if $(\mathcal{W}, w)$ is a minimizer of G with $g \in L^p(\Omega) \cap L^{np}_{\text{loc}}(\Omega)$ and $K = \overline{\Omega \setminus \bigcup \mathcal{W}}$; notice that K_1 is precisely of this type) there holds:

$$\limsup_{r \to 0} \omega_n^{-1} r^{-n} |V \cap B_{x,r}| \leq 1 - 8^{-n} \tag{24}$$

where V is any connected component of $\Omega \setminus K$ and $x \in \partial V \cap \Omega$. For, being V open and connected $\subset \Omega \setminus K = \Omega \cap \bigcup \mathcal{W}$, there will be $W \in \mathcal{W}$ s.t. $V \subset W$; were (24) false, we would find arbitrarily small $r > 0$ s.t. $|B_{x,r} \setminus W| < \omega_n 8^{-n} r^n$, and by the Clearing-out Lemma (with $m = 1$) a full neighborhood of x would be contained in W, hence $x \notin K$, contrary to the assumption $x \in \partial V \cap \Omega$.

We stress the fact that both estimates (22), (24) were derived under the assumption $g \in L^{np}_{\text{loc}}(\Omega)$. In this case we also have corresponding upper and lower area density estimates:

$$\limsup_{r \to 0} r^{1-n} \mathcal{H}^{n-1}(K \cap B_{x,r}) \leq n\omega_n \tag{25}$$

which holds for any optimal segmentation K and any $x \in \Omega$, and respectively:

$$\liminf_{r \to 0} r^{1-n} \mathcal{H}^{n-1}(K \cap B_{x,r}) > 0 \tag{26}$$

which holds for any segmentation K induced by an optimal partition: $K = \overline{\Omega \setminus \bigcup \mathcal{W}}$ with $(\mathcal{W}, w)$ minimizing G, and for any $x \in K \cap \Omega$; as for (23), it cannot hold for all minimizers of F.
The upper estimate (25) follows easily from (8) (since $P(\mathcal{W}, B_{x,r}) = \mathcal{H}^{n-1}(K \cap B_{x.r})$ if $\mathcal{W}$ is as in (21)), or more directly, from SPLIT & MERGE applied to K itself (i.e. by comparison of the minimizer (K,w) of F and the admissible pair (C,u) with $C = (K \setminus B_{x,r}) \cup \partial B_{x,r}$, $u = w$ on $\Omega \setminus \bar{B}_{x,r}$, $u = 0$ in $B_{x,r}$).

The lower estimate (26) can be proved by contradiction, using (as in the proof of (22)) the finiteness result established in Prop. 5, the Clearing-out Lemma, and the relative isoperimetric inequality, which implies in particular that

$$\begin{aligned} c(n) \cdot \min\{|B_{x,r} \cap W|, |B_{x,r} \backslash W|\}^{\frac{n-1}{n}} &\leq P(W, B_{x,r}) \\ &\leq \mathcal{H}^{n-1}(K \cap B_{x,r}) \end{aligned}$$

for all $W \in \mathcal{W}$.

When $g \in L^q_{\mathrm{loc}}(\Omega)$ for some $q > np$, we can considerably strengthen (26); in this case we have indeed

$$\lim_{r \to 0} r^{1-n} \mathcal{H}^{n-1}(K \cap B_{x,r}) \geq \omega_{n-1} \tag{27}$$

See [5], section 6, where (27) is shown in the case $g \in L^\infty(\Omega)$. In the same paper a regularity result is shown, still for a bounded datum g: this can also be extended to more general functions g, as we now show.

Let then $(\mathcal{W}, w)$ be a minimizer of G, with $g \in L^q_{\mathrm{loc}}(\Omega)$, $q > np$. Fix an arrangement $\{W_i\}$ of $\mathcal{W}$ and a point in $W_1(1/2) \cap W_2(1/2) \cap \Omega$, and as usual assume for simplicity it is the origin of $\mathbb{R}^n$. By the Clearing-out Lemma (with $m = 2$) we find a small ball $B_r \subset \Omega$ where $\mathcal{W}$ consists of just two sets, i.e. $(\mathcal{W}_r = \{W_1, W_2, \emptyset\}, w_r = w_{|B_r}) \in WCP(B_r)$ and $G_g(\mathcal{W}_r, w_r; B_r) \leq G_g(\mathcal{U}, u; B_r)$ for all compact variations $(\mathcal{U}, u)$ of $(\mathcal{W}_r, w_r)$ in B_r. In particular, taking $\mathcal{U} = \{U_1, U_2, \emptyset\}$ with $U_i = W_i$ outside $\bar{B}_{x,\rho} \subset B_r$, and $u(U_i) = w(W_i)$ $(i = 1, 2)$, we obtain easily

$$P(W_i, B_{x,\rho}) \leq P(U_i, B_{x,\rho}) + c \cdot \rho^{n-1+2\alpha} \quad (i = 1, 2)$$

with $\alpha = (q - np)/2q$. On the account of known regularity results for almost-minimal boundaries (see e.g. [9] or [10]), this yields

$$\partial^* W_1 \cap B_r = \partial^* W_2 \cap B_r \text{ is a hypersurface of class } C^{1,\alpha} \tag{28}$$

and

$$\mathcal{H}^{n-8+\varepsilon}(B_r \cap \partial W_i \backslash \partial^* W_i) = 0 \quad \forall \varepsilon > 0, i = 1, 2 \tag{29}$$

where $\partial^* W$ denotes the "reduced boundary" of W ([4], Chapter 3). As a corollary, we have that the interface $W_i(1/2) \cap W_j(1/2) \cap \Omega$ is (for $i \neq j$) a regular hypersurface of class $C^{1,\alpha}$ if $n \leq 7$, while, if $n \geq 8$, it is "mostly" such, with possible singularities of Hausdorff dimension not exceeding $n - 8$ (a well-know fact already in the context of interfaces of least area).

According to the Structure Theorem 3, we have

$$\Omega = \left(\bigcup_{i=1}^{\infty} R_i \right) \cup \left(\bigcup_{i<j} S_{ij} \right) \cup N \tag{30}$$

where the first sum concerns the "solid regions" $R_i = W_i \cap \Omega$ (which are open sets, recall Prop. 2) and the second one corresponds to the "interfaces" $S_{ij} = W_i(1/2) \cap W_j(1/2) \cap \Omega$ (to which the above "partial regularity result" applies).

The last term includes e.g. multiple singular points (for example, the vertex of the "standard plane trisector" of 120° angles); recall that $\mathcal{H}^{n-1}(N) = 0$.
For $i < j$ we have on the account of (28), (29):

$$S_{ij} = S_{ij}^* \cup N_{ij} \tag{31}$$

where $S_{ij}^* = \partial^* W_i \cap \partial^* W_j \cap \Omega$ is the regular part of the interface S_{ij} and $N_{ij} = S_{ij} \backslash S_{ij}^*$ is the singular one, with $\mathcal{H}^{n-8+\varepsilon}(N_{ij}) = 0 \; \forall \varepsilon > 0$. This holds whenever $\mathcal{W} = \{W_i\}$ is an optimal partition with $g \in L^p(\Omega) \cap L^q_{\text{loc}}(\Omega)$, $q > np$.
In terms of the associated optimal segmentation $K = \overline{\Omega \backslash \bigcup \mathcal{W}}$ we thus have

$$K \cap \Omega = \Omega \backslash \bigcup_{i=1}^{\infty} R_i = K_{\text{reg}} \cup K_{\text{sing}} \tag{32}$$

where $K_{\text{reg}} = \bigcup_{i<j} S_{ij}^*$ is a hypersurface of class $C^{1,(q-np)/2q}$ in Ω and K_{sing} is a closed subset of Ω with $\mathcal{H}^{n-1}(K_{\text{sing}}) = 0$.

To sum up, we can state the following

THEOREM 4 *Let Ω be an open subset of $\mathbb{R}^n$ $(n \geq 2)$, $\lambda > 0$, $p \geq 1$, $g \in L^p(\Omega)$, and consider the functional*

$$F(C, u) = \lambda \int_{\Omega \backslash C} |u - g|^p dx + \mathcal{H}^{n-1}(C \cap \Omega)$$

defined for C closed $\subset \mathbb{R}^n$ and $u \in C^1(\Omega \backslash C)$ s.t. $\nabla u(x) = 0 \; \forall x \in \Omega \backslash C$ (i.e., u is constant on each connected component of $\Omega \backslash C$).
If $g \in L^{np}_{\text{loc}}(\Omega)$, then F achieves its minimum, i.e. there exists (K, w) s.t. $F(K, w) \leq F(C, u)$ for all (C, u) as above. Moreover, the family of connected components of $\Omega \backslash K$ is locally finite in Ω, and the following estimates hold:

$$\liminf_{r \to 0} \omega_n^{-1} r^{-n} |V \cap B_{x,r}| > 0$$

for any connected component V of $\Omega \backslash K$ and all $x \in \partial V \cap \Omega$;

$$\limsup_{r \to 0} r^{1-n} \mathcal{H}^{n-1}(K \cap B_{x,r}) \leq n\omega_n \; \forall x \in \Omega.$$

Furthermore, given any minimizer (K, w) of F, a new minimizer (K_1, w_1) can be found s.t.

$$K_1 \subset K, \; K_1 = \overline{K_1 \cap \Omega}, \; H^{n-1}((K \backslash K_1) \cap \Omega) = 0;$$

$$w_1 = w \text{ in } \Omega \backslash K;$$

$$\limsup_{r \to 0} \omega_n^{-1} r^{-n} |V \cap B_{x,r}| \leq 1 - 8^{-n}$$

for any connected component V of $\Omega \backslash K_1$ and all $x \in \partial V \cap \Omega$;

$$\liminf_{r \to 0} r^{1-n} \mathcal{H}^{n-1}(K_1 \cap B_{x,r}) > 0 \; \forall x \in K_1 \cap \Omega.$$

If $g \in L^q_{\mathrm{loc}}(\Omega)$ *with* $q > np$ *then we actually have*

$$\lim_{r \to 0} r^{1-n} \mathcal{H}^{n-1}(K_1 \cap B_{x,r}) \geq \omega_{n-1} \, \forall x \in K_1 \cap \Omega.$$

In this case moreover $K \cap \Omega = K_{\mathrm{reg}} \cup K_{\mathrm{sing}}$, *where* K_{reg} *is a smooth hypersurface of class* $C^{1,(q-np)/2q}$ *in* Ω *and* K_{sing} *is a closed subset of* Ω *with* $\mathcal{H}^{n-1}(K_{\mathrm{sing}}) = 0$.

It seems reasonable that in fact $\mathcal{H}^{n-2}(K_{\mathrm{sing}} \cap A) < \infty \quad \forall A \subset\subset \Omega$: this is easily seen to be true for $n = 2$, by using the special structure of 2-dimensional singularities ("triple points"; see also [8], Lemma 5.7).

We also notice that some of the preceding results can be extended "up to the boundary" of Ω: the interested reader is referred to [6] for the counterparts of Lemma 1 and Prop. 4, which are the key tools for the proof that (under appropriate assumptions on Ω and g) optimal partitions are globally finite.

References

[1] **G. Congedo, I. Tamanini,** *On the existence of solutions to a problem in multidimensional segmentation*, Ann. Inst. Henri Poincaré **8** (1991), 175–195.

[2] **E. De Giorgi,** *Free discontinuity problems in calculus of variations,* in: *Frontiers in pure and applied mathematics,* R.Dautray Editor, North-Holland, Amsterdam (1991)

[3] **E. De Giorgi, M. Carriero, A. Leaci,** *Existence theorem for a minimum problem with free discontinuity set*, Arch. Rat. Mech. Anal. **108** (1989), 195–218.

[4] **E. Giusti,** *Minimal surfaces and functions of bounded variation*, Birkhäuser, Basel (1984).

[5] **U. Massari, I. Tamanini,** *Regularity properties of optimal segmentations*, J. reine angew. Math. **420** (1991), 61–84.

[6] **U. Massari, I. Tamanini,** *On the finiteness of optimal partitions,* Ann. Univ. Ferrara **39** (1994), to appear.

[7] **J.M. Morel, S. Solimini,** *Variational Methods in Image Segmentation,* Birkhäuser, Basel (1994).

[8] **D. Mumford, J. Shah,** *Optimal approximations by piecewise smooth functions and associated variational problems*, Comm. Pure Appl. Math. **42** (1989), 577–685.

[9] **I. Tamanini**, *Boundaries of Caccioppoli sets with Hölder-continuous normal vector,* J. reine angew. Math. **334** (1982), 27–39.

[10] **I. Tamanini,** *Regularity results for almost minimal oriented hypersurfaces in* $\mathbb{R}^n$, Quaderno Dip. Mat. Univ. Lecce **1** (1984).

[11] **I. Tamanini, G. Congedo,** *Optimal segmentation of unbounded functions*, to appear in Rend. Sem. Mat. Univ. Padova (1995).

Progress in Nonlinear Differential Equations
and Their Applications, Vol. 25

Indefinite Superlinear Elliptic Problems

Italo Capuzzo Dolcetta

Dipartimento di Matematica, Università "La Sapienza"
P. le A. Moro 2, 00185 Roma, Italy

1. Introduction

This presentation is aimed at giving a rather complete description of *existence* and *nonexistence* of *positive* solutions of the following semilinear elliptic equation

$$\begin{cases} -\Delta u + q(x)u = \tau u + a(x)u^p, & \text{in } \Omega \\ Bu = 0 & \text{on } \partial\Omega \end{cases} \tag{1.1}$$

where τ is a real parameter and q is a given continuous function. The boundary operator B denotes either one of the following boundary conditions on $\partial\Omega$:

$$Bu := \partial_\nu u = 0, \text{ or } Bu := u = 0, \text{ or } Bu := \partial_\nu u + \alpha(x)u = 0 \tag{1.2}$$

where $\alpha > 0$ is a given continuous function on $\partial\Omega$.
Here, Ω is a bounded (in 5 we will also consider the case of unbounded domains) smooth domain of $\mathbb{R}^N$, ν is the outward unit normal, $a(x)$ is a continuous function which *may change sign* in Ω and $p > 1$ which makes the problem *superlinear.*
A principal feature of problem (1.1) is the interplay between its *indefinite character*, due to the changing of sign of a, and the spectral properties of the linear operator on the left hand side.

Quite recently there has been some attention given to superlinear indefinite problems. We refer the reader to [14] and [11] for Hamiltonian systems. For elliptic nonlinear problems some results have been obtained in the sublinear case and for Neumann boundary condiitions in [3].
Superlinear problems like (1.1) have been studied in [15] via bifurcation methods and in [1] by variational methods. The most complete results concerning problem (1.1) are due to H. Berestycki, L. Nirenberg and the author [4, 5]. They form the core of this presentation. Let us mention also that more general nonlinearities have been treated in [6] by topological methods.
Some *degenerate elliptic* cases have been studied in [8, 7]; the main results from those papers are reported here.

The plan of presentation is as follows:

2. A nonlinear Picone identity and necessary conditions.
3. Existence results.
4. A superlinear problem for the Heisenberg Laplacian.
5. Non existence results in unbounded domains: Liouville theorems.

2. A nonlinear Picone identity and necessary conditions

In this section we present necessary conditions for the existence of positive solution of

$$\begin{cases} u > 0 & \\ -\Delta u + q(x)u = \tau u + a(x)u^p, & \text{in } \Omega \\ Bu = 0 & \text{on } \partial\Omega \end{cases} \tag{2.3}$$

We assume that the linear operator $-\Delta + q(x)$ with the boundary condition B has zero as its principal eigenvalue. That is, we make the following assumption on q:

$$\begin{cases} \text{there exists } \Phi > 0 \text{ such that} \\ -\Delta\Phi + q(x)\Phi = 0 \ \text{ in } \Omega, \ \ B\Phi = 0 \text{ on } \partial\Omega. \end{cases} \tag{2.4}$$

Under this assumption, necessary conditions for the solvability of problem (2.3) depending on the values of τ are given by the next:

THEOREM 2.1 *Assume that (2.4) holds and that (2.3) has a solution. Then, necessarily,*

(i) $\int_\Omega a\Phi^{p+1}dx < 0$, *if* $\tau > 0$,

(ii) $\Omega^+ := \{x \in \Omega; a(x) > 0\} \neq \emptyset$, *if* $\tau < 0$,

(iii) $\Omega^+ \neq \emptyset$ *and* $\int_\Omega a\Phi^{p+1}dx < 0$ *if* $\tau = 0$.

The proof of Theorem 2.1 is based on the following *nonlinear Picone identity*

$$div\left[\psi\left(\frac{\phi}{u}\right)(u\nabla\phi - \phi\nabla u)\right] = \psi\left(\frac{\phi}{u}\right)(u\Delta\phi - \phi\Delta u) + \psi'\left(\frac{\phi}{u}\right)u^2\left|\nabla\left(\frac{\phi}{u}\right)\right|^2 \tag{2.5}$$

which holds for any $\psi \in C^1(\mathbb{R})$, $u, \phi \in C^2$, $u > 0$. For $\psi(t) = t$ the above reduces to the classical Picone identity (see [13] for a review of some of its applications).

In the proof of theorem 2.1 the above identity is used with $\psi(t) = t^\gamma$ with suitable γ, u and Φ satisfying (2.3) and (2.4), respectively (see [4]).

Let us mention that in the special case $q = 0$ and $\tau = 0$ the necessary conditions (iii) were derived in [5]. For the case of Dirichlet boundary conditions with $q = -\lambda_1$, the principal eigenvalue of $-\Delta$, the necessity of (i) was established (by a different method) in [1].

3. Existence results

Let us consider first the case where a is of *constant sign* in Ω. In this simpler situation one obtains necessary and sufficient conditions. Namely, we have

THEOREM 3.1 *If* $a < 0$ *in* Ω, *then problem (2.3) has a solution if and only if* $\tau > 0$. *If* $a > 0$ *in* Ω, *then problem (2.3) has a solution if and only if* $\tau < 0$.

The existence of solutions follows from variational methods like the Mountain Pass Lemma (see [2, 16]) in the case $a > 0$. The case $a < 0$ is dealt with using sub-supersolution techniques in the sublinear case $p < 1$.

In the case where a *changes the sign* in Ω the situation is more complicated and the standard approach of critical point theory does not apply, at least in the *non-coercive* case $\tau \geq 0$. In this case, a method combining constrained maximization and sub-supersolution arguments allows to establish (see [5]) the following:

THEOREM 3.2 *Assume (2.4),* $1 < p < \frac{N+2}{N-2}$, *and*

(i) $\Omega^+ \neq \emptyset$ *and* $\Omega \setminus \Omega^+ \neq \emptyset$.

(ii) $\int_\Omega a\Phi^{p+1}dx < 0$.

Then there exists τ^* *such that problem (2.3) has a solution for* $0 \leq \tau < \tau^*$, *while no solution exists for* $\tau > \tau^*$.

Let us point out here that the question of multiplicity of solutions is *open* in general. However, some results in this direction are known (see [17] for Neumann conditions, [1] for Dirichlet conditions with $q = -\lambda_1$).

For the *coercive* case $\tau < 0$, it is not hard to check that the Palais-Smale condition and the geometrical assumptions of the Mountain Pass Lemma are fulfilled by the functional

$$J(u) = \int_\Omega \left\{ \frac{1}{2}|\nabla u|^2 + \frac{(q(x) - \tau)}{2}u^2 - \frac{a(x)(u^+)^{p+1}}{p+1} \right\} dx$$

if and only if a is strictly positive at some point in Ω (see [5] and also, for the Dirichlet case, [2, 16]). Hence, we have:

THEOREM 3.3 *For* $\tau < 0$ *problem (2.3) has a solution if and only if* a *is positive at some point of* Ω.

4. A superlinear problem for the Heisenberg Laplacian

Consider now the Dirichlet problem

$$\begin{cases} u > 0, \quad -\Delta_H u + q(\xi)u = \tau u + a(\xi)u^p, & \text{in } \Omega \\ u = 0 & \text{on } \partial\Omega \end{cases} \tag{4.6}$$

Here, Ω is a bounded smooth domain of $\mathbb{R}^{2n+1}$ whose generic point is denoted by $\xi := (x_1, .., x_n, y_1, .., y_n, t)$, and Δ_H is the Heisenberg Laplacian, i.e.

$$\Delta_H = \sum_{i=1}^n \frac{\partial^2}{\partial x_i^2} + \frac{\partial^2}{\partial y_i^2} + 4(x_i^2 + y_i^2)\frac{\partial^2}{\partial t^2} + 4y_i\frac{\partial^2}{\partial x_i \partial t} - 4x_i\frac{\partial^2}{\partial y_i \partial t}.$$

Existence and nonexistence for semilinear equation involving the *degenerate elliptic* operator Δ_H have been studied in [9]; their results cover the case $a > 0$ in Ω.
The case of indefinite nonlinearities has been considered in [8]. Under the spectral assumption

$$\begin{cases} \text{there exist } \Phi > 0 \text{ and } \tau_1 > 0 \text{ such that:} & \\ -\Delta_H \Phi + q(\xi)\Phi = \tau_1 \Phi & \text{in } \Omega \\ \Phi = 0 & \text{on } \partial\Omega \end{cases} \tag{4.7}$$

necessary conditions analogous to those in Theorem 2.1 are obtained using the fact that Δ_H satisfies the Hörmander condition of order one, see [12].

Concerning existence of solutions in the *non coercive* case $\tau \geq \tau_1$ they prove the following:

THEOREM 4.1 *Assume (4.7), $1 < p < \frac{2n+4}{2n}$ and $\Omega^+ \neq \emptyset$, $\int_\Omega a\Phi^{p+1} d\xi < 0$. Then there exists τ^* such that (4.6) has a solution for $\tau_1 \leq \tau < \tau^*$, while no solution exists for $\tau > \tau^*$.*

Observe that in the present degenerate elliptic case the growth of the nonlinearity is restricted by the critical exponent $\frac{Q+2}{Q-2}$, where $Q = 2n+2$ is the homogeneous dimension of $\mathbb{R}^{2n+1}$ endowed with the Heisenberg group action

$$\tilde{\xi} \circ \xi = (x + \tilde{x}, y + \tilde{y}, t + \tilde{t} + 2\sum_{i=1}^{n}(x_i \tilde{y}_i - y_i \tilde{x}_i)).$$

The variational methods used in [5] are applied in the Sobolev-Stein space $\overset{o}{S}{}^2_1$; some typical difficulties are overcome through a suitable version of Hopf's lemma.

5. Non-existence results in unbounded domains: Liouville theorems

In this section we consider the equation

$$-\Delta u = a(x) u^p \text{ in } \Sigma \tag{5.8}$$

where Σ is an open connected cone in $\mathbb{R}^N$ $(N \geq 2)$ with vertex at the origin. For the case $\Sigma = \mathbb{R}^N$ and $a(x) \equiv A > 0$, it is known ([10]) that, for $1 < p < \frac{N+2}{N-2}$, if u is a smooth solution of (5.8), then necessarily $u \equiv 0$. Assume more generally,

$$\begin{cases} \bar{\Sigma} \neq \mathbb{R}^N, a \geq 0, \ a \text{ is continuous in } \Sigma \text{ and bounded near } 0 \\ a(x) = A\,|x|^\gamma \ \text{ for } |x| \text{ large}, \gamma > -2, A > 0. \end{cases} \tag{5.9}$$

Let λ_1 be the principal eigenvalue for the Dirichlet problem on $\Sigma \cap S^{N-1}$ of the Laplace-Beltrami operator on S^{N-1} and define α by the identity

$$\lambda_1 = \alpha(N - 2 + \alpha).$$

The following Liouville-type theorem has been established in [6]:

THEOREM 5.1 *Assume (5.9),* $1 < p < \frac{N+\alpha+\gamma}{N-2+\alpha}$*. If* $u \in C^2(\Sigma)$ *is a smooth solution of (5.8), bounded near the origin, then* $u \equiv 0$.

The result applies, in particular, to the equation

$$-\Delta u = x_N u^p \tag{5.10}$$

in the half space $\Sigma = \{x \in \mathbb{R}^N : x_N > 0\}$. By (5.1), its unique nonnegative solution which is bounded near the origin is $u \equiv 0$, provided that $1 < p < \frac{N+2}{N-1}$.

Rather surprisingly, similar Liouville properties hold when Δ is replaced by Δ_H. Let us just mention here, among the results of [7], the following analogue of the one in [10] mentioned at the beginning of this section.

THEOREM 5.2 *Assume* $1 < p < \frac{Q}{Q-2}$ *and* $A > 0$ *. If* u *is a smooth nonnegative solution of*

$$-\Delta_H u = A u^p \ \text{in} \ \mathbb{R}^{2n+1}, \tag{5.11}$$

then $u \equiv 0$.

References

[1] S. ALAMA, G. TARANTELLO, On semilinear elliptic equations with indefinite nonlinearities *Calc. of Var. and PDE* **1** (1993) 439–475.

[2] A. AMBROSETTI, P. RABINOWITZ, Variational methods in critical point theory and applications *J. Funct. Anal.* **14** (1973) 349–381.

[3] C. BANDLE, M.A. POZIO, A. TESEI, Existence and uniqueness of solutions of nonlinear problems, *Math. Z.* **199** (1988) 257–278.

[4] H. BERESTYCKI, I. CAPUZZO DOLCETTA, L. NIRENBERG, Problèmes elliptiques indéfinis et théorèmes de Liouville non-linéaires, *C. R. Acad. Sci. Paris, Série I* **317** (1993) 945–950.

[5] H. BERESTYCKI, I. CAPUZZO DOLCETTA, L. NIRENBERG, Variational methods for indefinite superlinear homogeneous elliptic problems, to appear in *Nonlinear Differential Equations and Applications.*

[6] H. BERESTYCKI, I. CAPUZZO DOLCETTA, L. NIRENBERG, Superlinear indefinite elliptic problems and nonlinear Liouville theorems, *Topological Methods in Nonlinear Analysis* **4** (1994) 59–78.

[7] I. BIRINDELLI, I. CAPUZZO DOLCETTA, A. CUTRI, Liouville theorems for semilinear equations on the Heisenberg group, to appear in *Annales IHP, Analyse Nonlineaire* .

[8] I. BIRINDELLI, A. CUTRI, A semilinear problem for the Heisenberg Laplacian, to appear in *Rendiconti Sem. Mat. Univ. Padova* .

[9] N. GAROFALO, E. LANCONELLI, Existence and non existence results for semilinear Equations on the Heisenberg Group, *Indiana Univ. Math. Journ.***41** (1992) 71–97.

[10] B. GIDAS, J. SPRUCK, Global and local behavior of positive solutions of nonlinear elliptic equations, *Comm.Pure Appl. Math.* **35** (1981) 525–598.

[11] M. GIRARDI, M. MATZEU, Existence and multiplicity results for periodic solutions of superquadratic Hamiltonian systems where the potential changes the sign, preprint Univ. di Roma "Tor Vergata", 7–93.

[12] L. HÖRMANDER, Hypoelliptic second order differential equations, *Acta Math.* **119** (1967) 147–171.

[13] K. KREITH, Picone identity and generalizations, *Rendiconti di Matematica* **8** (1975) 251–262.

[14] S. LASSOUED, Periodic solutions of a second order superquadratic system with change of sign of potential, *J. Diff. Eq.* **93** (1991) 1–18.

[15] T. OUYANG, On the positive solutions of semilinear equation $\Delta u + \lambda u + hu^p = 0$ on compact manifolds, Part II, *Indiana Univ. Math. J.* **40** (1991) 1083–1141.

[16] M. STRUWE, *Variational methods and applications to nonlinear partial differential equations and Hamiltonian systems*, Springer Verlag (1990)

[17] H. TEHRANI, Remarks on some semilinear elliptic eigenvalue problems, PhD thesis, Courant Institute NYU, (1994).

Progress in Nonlinear Differential Equations
and Their Applications, Vol. 25

On the Regularity of the Edge Set of Mumford-Shah Minimizers

ALEXIS BONNET

University of Cergy-Pontoise 8, le Campus, 95033 Cergy-Pontoise Cedex

ABSTRACT. We present a series of results on the regularity of edges in the Mumford-Shah minimization process. An open problem is whether the edge of a minimizer is made of a finite number or arcs. A partial answer to this question is given in this article.

1. Introduction

In computer vision, an image is defined by the grey-level at each point of the screen. We denote by g the function which associates to each point of the image its grey-level. We will assume that g is an L^∞ function on a domain Ω of $\mathbb{R}^2$ and takes its values in the interval $[0, 1]$. At a first glance, an image appears as made of patches inside which the grey-level has but little variations. Sharp discontinuities are introduced by edges of objects, shadows, overlapping objects, etc. Looking closer, other small discontinuities or defects may appear. It is natural to expect that a regularized image should keep the meaningful discontinuities. The object of image segmentation is to find a piecewise smooth function u approximating the true image g and such that the set K of discontinuities of u is of minimal length.

Mumford and Shah introduced (u, K) in [17] as minimizers of the functional:

$$J(u,K) = \int_{\Omega\setminus K} ||\nabla u||^2 + \mu^2 \int_\Omega (u-g)^2 + \nu\mathcal{H}^1(K). \qquad (1)$$

This energy is made of three terms. The first one penalizes the variations of u outside the edge set K, the second one is the distance to the true image and the third one is the 1-dimensional Hausdorff measure of the edge set K [15] and [12]. In the energy J, μ and ν denotes two positive constants. By a renormalization we may assume without loss of generality that $\mu = \nu = 1$. Mumford and Shah conjectured that there exits a minimizer of J such that the edge set K is the union of a finite set of C^1-arcs and that each arc may end either as a crack-tip or in a triple junction.

Existence of a minimizer in the class of SBV functionals has been proved simultaneously in [2], [11] and [7]. It is proved in [7] that there is a correspondence between SBV minimizers and pairs (u, K) minimizing J. In all the following we will work in the (u, K) formulation.

Notice that we do not change $J(u, K)$ by adding a set of zero $\mathcal{H}^1$ measure to K. Therefore, we will assume in the following that K is a minimal closed set:

there is no closed set $\tilde{K} \subset K$, $\tilde{K} \neq K$, such that u can be extended on $\Omega\backslash\tilde{K}$ with $J(u,\tilde{K}) \leq J(u,K)$.

An interesting property of minimizers of J has been proved in [7]: there are two positive constants c and C such that

$$\text{if } P \in K \text{ then } \forall r > 0,\ cr \leq \mathcal{H}^1(K \cap B_r(P)) \leq Cr. \tag{2}$$

In this paper, we are interested in the local regularity of the edge set (see also [9] and [3]-[4]) and in the number of arcs. Our main ingredient is the characterization of all global-local minimizers (i.e. pairs (u, K) defined on $\Omega = \mathbb{R}^2$ and such that for all balls B_R the energy $J^0_{B_R}(u,K) = \int_{B_R\backslash K} ||\nabla u||^2 + \mathcal{H}^1(K \cap B_R)$ cannot be reduced by a local modification of u and K inside the ball B_R). Under the assumption that K is connected, it will be proved that there are only four types of global-local minimizers. The characterization is carried out by a monotonicity property on the function $r \mapsto \frac{\int_{B_r\backslash K} ||\nabla u||^2}{r}$. We will be able to prove that every blow-up limit is a global-local-minimizer. The characterization of all global-local-minimizers and therefore of the blow-up limits give strong information on the local behavior of K and u.

We give below a simple formulation of the main results:

THEOREM 1.1 *If G is an isolated connected component of K then it is the union of a finite set of C^1 arcs. These arcs are $C^{1,1}$ away from crack-tips and can merge only through triple junctions.*

More precisely:

THEOREM 1.2 *Let P be a point of the edge K, if $B_R(P)$ intersects one connected component of K, then there is a finite set of arcs l_j $j = 1,..,k$ with the same regularity property as above and such that*

$$K \cap B_{R/2}(P) = \left(\cup_{j=1}^k l_j\right) \cap B_{R/2}(P). \tag{3}$$

Notice that this does not allow a minimizer to have an infinite number of arbitrary small pieces connected to each other. This is a step towards the finite number of C^1 arcs in the Mumford-Shah conjecture. In July 1994, G. David has announced a $\mathcal{H}^1$ almost everywhere regularity result, [9]. Actually the method used to prove the theorems above allows one to get

THEOREM 1.3 *For $\mathcal{H}^1$ almost every P in K there is a neighborhood $B_r(P)$ of P in which K is a $C^{1,1}$ arc.*

The above results were derived when g is L^∞. If one assumes more regularity on g one may expect more regularity on K.

THEOREM 1.4 *Assume that g is $C^{k,\alpha}$. If, in an open neighborhood U of $P \in K$, $K \cap U$ is a C^1 arc then it is $C^{k+2,\alpha}$ in U.*

Lastly we would like to point out that it can be interesting to look for a minimizer (u^n, K^n) of J that has at most n connected components (see [15]) We have the complete result:

THEOREM 1.5 *The edge set K^n is the union of a finite set of C^1 arcs, $C^{1,1}$ away from the crack-tips, and merging through triple junctions.*

Open questions. It is not known whether the edge set is regular up to the crack-tip. The finite number of arcs in the most general setting is still an open question. Notice that if the characterization of global-local-minimizers was achieved without the assumption that K is connected then the conjecture would be solved.

We present a brief outline of the proof of the results (the reader may refer to [5] for the details).

2. Blow-up procedures

Blow-up procedures are common in the study of free-boundary problems ([1], [14]). Here, we introduce, for a given origin, the blow-up sequences u_ϵ, g_ϵ, K_ϵ:

$$\begin{aligned} u_\epsilon(X) &= \tfrac{u(\epsilon X) - c_\epsilon(X)}{\sqrt{\epsilon}}, \\ g_\epsilon(X) &= \tfrac{g(\epsilon X) - c_\epsilon(X)}{\sqrt{\epsilon}}, \\ K_\epsilon &= \{X \in \mathbb{R}^2, \epsilon X \in K\}. \end{aligned} \tag{4}$$

The real function $c_\epsilon(X)$ is piecewise constant. It is introduced in order to be able to pass to the limit $\epsilon \longrightarrow 0$ such that u_ϵ converges almost everywhere to some finite value. We first take $c \equiv 0$ and introduce:

$$\begin{aligned} U_\epsilon(X) &= \tfrac{u(\epsilon X)}{\sqrt{\epsilon}}, \\ G_\epsilon(X) &= \tfrac{g(\epsilon X)}{\sqrt{\epsilon}}. \end{aligned} \tag{5}$$

We notice that $\nabla U_\epsilon(X) = \sqrt{\epsilon} \nabla u(\epsilon X)$ and that a ball B_ϵ of radius ϵ is sent on B_1 by the blow-up procedure. Let J_ω denote the energy $J_\omega(v, \tilde{K}) = \int_{\omega \setminus \tilde{K}} |\nabla v|^2 + \int_\omega (v-g)^2 + \mathcal{H}^1(\tilde{K} \cap \omega)$, we have then

$$J_{B_\epsilon}(u, K) = \epsilon \int_{B_1 \setminus K_\epsilon} |\nabla U_\epsilon|^2 + \epsilon^3 \int_{B_1} (U_\epsilon - G_\epsilon)^2 + \epsilon \mathcal{H}^1(K_\epsilon \cap B_1). \tag{6}$$

We introduce the new functionals J_ω^ϵ:

$$J_\omega^\epsilon(U_\epsilon, K_\epsilon) = \int_{\omega \setminus K_\epsilon} |\nabla U_\epsilon|^2 + \epsilon^2 \int_\omega (U_\epsilon - G_\epsilon)^2 + \mathcal{H}^1(K_\epsilon \cap \omega). \tag{7}$$

With this definition we have the straightforward statement on the minimization under Dirichlet conditions (minimize $J_\omega(v, G)$ with $v = u$ on $\partial\omega$):

$$(u, K) \text{ minimizes } J_{B_\epsilon} \Leftrightarrow (U_\epsilon, K_\epsilon) \text{ minimizes } J_{B_1}^\epsilon. \tag{8}$$

We remark that in the definition of J^ϵ_ω, the term $\epsilon^2 \int_\omega (U_\epsilon - G_\epsilon)^2$ is of order ϵ. In the limit $\epsilon = 0$ we define J^0_ω:

$$J^0_\omega(u_0, K_0) = \int_{\omega\backslash K_0} |\nabla u_0|^2 + \mathcal{H}^1(K_0 \cap \omega). \tag{9}$$

In the following we will be interested in minimizers of the functional J^0 that are defined on the whole plane. Since there is no more image g in the functional J^0 and since the domain is now unbounded, we need a new definition of a minimizer for J^0.

DEFINITION 2.1 *A segmentation (u, K) is a global-local minimizer of J^0 if it is defined on the whole space and if for any bounded open sub-domains U, V of $\mathbb{R}^2$ with $U \subset\subset V$ there is no function $\tilde{u}$ and edge set $\tilde{K}$ defined on V such that:*
(i) $\tilde{u}$ and u (resp. $\tilde{K}$ and K) coincide in $V\backslash\bar{U}$,
(ii) for every pair of points X, Y in $V\backslash(\bar{U} \cup K)$, if X and Y are in two distinct connected components of $V\backslash K$ so they are in $V\backslash\tilde{K}$,
(iii) $J^0_V(\tilde{u}, \tilde{K}) < J^0_V(u, K)$.

Condition (ii) is coherent with the introduction of the piecewise constant function $c_\epsilon(X)$ in the definition of the blow-up sequences. This ensures that if (u_0, K_0) be the segmentation of the plane such that K_0 is a straight line $\{y = 0\}$ and u is constant on both sides of K ($u_0(x, y) = c^+$ if $y > 0$, $u_0(x, y) = c^-$ if $y < 0$) then (u_0, K_0) is a global-local minimizer.

The fundamental result in the blow-up procedure is the following theorem which says that blow-up sequences converge to global-local minimizers.

THEOREM 2.2 *Let (u, K) be a minimal segmentation of J. The blow-up sequence u_ϵ; K_ϵ is such that there is a subsequence converging to u_0, K_0 in the following sense:*

$$\begin{gathered} K_{\epsilon_q} \to K_0 \text{ locally in the Hausdorff metric,} \\ u_{\epsilon_q} \to u_0 \text{ strongly in } H^1_{loc}(\mathbb{R}^2\backslash K_0), \end{gathered} \tag{10}$$

$$J^0_\omega(u_0, K_0) \leq \liminf J^{\epsilon_q}_\omega(U_{\epsilon_q}, K_{\epsilon_q}) \text{ for any bounded open set } \omega,$$

$$J^0_{B_r}(u_0, K_0) = \lim J^{\epsilon_q}_{B_r}(U_{\epsilon_q}, K_{\epsilon_q}) \text{ for almost every } r > 0. \tag{11}$$

Moreover, (u_0, K_0) is a global-local minimizer.

Proof. The proof of convergence uses the compactness of the Hausdorff distance for the sequence K_ϵ and the uniform bounds on ∇U_ϵ. The limit is a global-local minimizer thanks to a lower semicontinuity lemma of [7] and a construction which is detailed in [5].

3. Monotonicity formulas

In this section we state a few monotonicity formulas which will be essential for the characterization of all global-local-minimizers. We consider a global-local-minimizer (u, K) and introduce $\phi(r) = \int_{B_r\backslash K} |\nabla u|^2$.

THEOREM 3.1 *If K is connected then $r \mapsto \frac{\phi(r)}{r}$ is a non-decreasing function of r. If $\frac{\phi(r)}{r}$ is constant, then in some polar coordinates (r, θ) we have:*

$$u(r,\theta) = C\sqrt{r}\cos\frac{\theta}{2} \text{ for } \theta \in [0, 2\pi)$$

and K is the half axis $\{\theta = 0\}$.

Monotonicity formulas of an other type have been used in the study of two-phase free-boundary problems (see for instance [1]).

For the study of the edge set near the boundary of the domain, we consider global-local minimizers defined in a half-space $\mathbb{R}^+ \times \mathbb{R}$. In this case, we have

COROLLARY 3.2 *If (u, K) is a global-local minimizer in $\mathbb{R}^+ \times \mathbb{R}$ such that $K \cup \partial(\mathbb{R}^+ \times \mathbb{R})$ is connected then $r \mapsto \frac{\phi(r)}{r^2}$ is a non-decreasing function of r.*

REMARK 3.3 *In fact, if U is a Lipschitz domain and if (u, K) is a global-local minimizer in U such that $K \cup \partial U$ is connected then there is a $\alpha \in (1, 2]$ such that $r \mapsto \frac{\phi(r)}{r^\alpha}$ is a non-decreasing function of r.*

4. Characterization of global-local minimizers

The main theorem of this section is:

THEOREM 4.1 *If (u, K) is a global-local minimizer such that K is connected then (u, K) is one of the following:*

(i) K is empty and u is constant.

(ii) K is a straight line defining two half-planes and u is constant on each half-plane.

(iii) K is the union of three half lines with angle $2\pi/3$ and u is constant on each sector.

(iv) in a polar set of coordinates $u(r,\theta) = \sqrt{\frac{2}{\pi}}\sqrt{r}\cos\frac{\theta}{2}$ for $\theta \in [0, 2\pi)$ and K is the half axis $\theta = 0$.

Proof. The determination of the constant $\sqrt{\frac{2}{\pi}}$ can be found in [17].

The proof of the theorem relies on a careful use of the monotonicity formula. It will be completed once we have shown that $\frac{\phi(r)}{r}$ is a constant for a properly chosen origin. We start by studying the limits: $\lim_{+\infty} \frac{\phi(r)}{r}$ and $\lim_0 \frac{\phi(r)}{r}$ using blow-up and blow-down technic. The blow-up is defined as in section 2. and the blow-down is given by:

$$u_l(X) = \frac{u(lX)}{\sqrt{l}}, \quad K_l = \{X \in \mathbb{R}^2, lX \in K\} \tag{12}$$

where $l \to +\infty$. From theorem 2.2 we deduce the convergence of blow-up and blow-down sequences to global-local-minimizers (u_0, K_0) and (u_∞, K_∞) respectively.

LEMMA 4.2 *Let* $\phi_0(r) = \int_{B_r} |\nabla u_0|^2$ *and* $\phi_\infty(r) = \int_{B_r} |\nabla u_\infty|^2$. *If* K *is connected then*

$$\frac{\phi_0(r)}{r} \equiv \lim_0 \frac{\phi(\rho)}{\rho} \text{ and } \frac{\phi_\infty(r)}{r} \equiv \lim_\infty \frac{\phi(\rho)}{\rho}. \tag{13}$$

Theorem 3.1 with $C = \sqrt{\frac{2}{\pi}}$ implies that $\lim_0 \frac{\phi(r)}{r}$ and $\lim_\infty \frac{\phi(r)}{r}$ can only take the value 0 or 1. If $\lim_\infty \frac{\phi(r)}{r} = 0$, the monotonicity formula gives $\phi(r) \equiv 0$. Consequently, u is a constant in any connected component of $\mathbb{R}^2 \backslash K_0$ and we are reduced to minimizing $\mathcal{H}^1(K_0)$ under the condition (ii) of definition 2.1 which gives situations (i), (ii) or (iii) of theorem 4.1.

The case where $\lim_\infty \frac{\phi(r)}{r} = 1$ is solved by the proposition:

PROPOSITION 4.3 *If* $\lim_\infty \frac{\phi(r)}{r} = 1$ *then there is a point* P *such that if we denote* $\phi_P(r) = \int_{B_R(P)} |\nabla u|^2$ *then:*

$$\lim_0 \frac{\phi_P(r)}{r} = 1. \tag{14}$$

We will say that P *is a crack-tip for the edge set* K.

Proof. This can be proved by a contradiction argument. The idea is to build a blow-up (or blow-down) limit such that there is a closed curve in K defining a bounded connected component in $\mathbb{R}^2 \backslash K$. We then conclude by:

LEMMA 4.4 *If* (u, K) *is a global-local minimum. Then* $\mathbb{R}^2 \backslash K$ *cannot have a bounded connected component.*

Proof. Notice that by an obvious excision argument any bounded connected component of $\mathbb{R}^2 \backslash K$ is convex. By contradiction assume that $\mathbb{R}^2 \backslash K$ has a bounded convex connected component $\mathcal{O}$. Now, u is necessary a constant in the bounded connected component $\mathcal{O}$. Let A be a regular point of $\partial\mathcal{O}$. The blow-up at A is a line (situation (ii)). There is a limit value of u on both side of K at A. We may modify u inside $\mathcal{O}$ such that those two values are identical. Then the blow-up at A can be proceeded with a function c constant in the whole plane. Therefore, the blow-up limit has to be a global-local minimum with respect to any local modification (the assumption (ii) of definition 2.1 is not required). Here, the blow-up limit (u_0, K_0) is such that K_0 is a line and $u_0 \equiv 0$ in $\mathbb{R}^2 \backslash K_0$. Consequently for any ball B_R one can take $\tilde{K}_0 = K_0 \backslash B_R$ and $\tilde{u} \equiv 0$ in $\mathbb{R}^2 \backslash \tilde{K}_0$ and reduce the J^0 energy. This contradicts the minimality of (u_0, K_0) and completes the proof of lemma 4.4.

5. Finite number of pieces

In this section we will prove that any isolated connected component of K is made of a finite number of curves.

THEOREM 5.1 *Let* Ω *be a Lipschitz domain and* G *be an isolated connected component of* K. *Then:*

$$G = \cup_{i=1}^k \psi_i([0,1]) \tag{15}$$

where ψ_i *is a one to one Lipschitz function and* $\psi_i((0,1)) \cap \psi_j((0,1)) = \emptyset$ *if* $i \neq j$.

Proof. With the

LEMMA 5.2 *Assume that $\omega \subset \Omega$ intersects only one connected component of K. Let P_n be a sequence of points of an open set $U \subset\subset \omega$. For any blow-up sequence $u_n(X) = \frac{u(P_n+\epsilon_n X)}{\sqrt{\epsilon_n}}$, K_n converging to a blow-up limit (u_0, K_0) the limit edge set K_0 is connected.*

we can use theorems 2.2 and 4.1 to conclude that any point is a flat point (ii) or a triple point (iii) or a crack-tip (iv). We have to prove then that there is a finite number of triple points. By contradiction, assume that there is a sequence T_n of distinct triple points in G. By compactness we may assume that T_n converges to a point P of G. An appropriate blow-up sequence will converge to a blow-up limit (u_0, K_0) such that K_0 is connected but which cannot be (by construction) of any of the types of theorem 4.1. With the finite number of triple points then one can prove that G is a finite union of pairwise disjoint rectifiable Jordan arcs.

6. Regularity $C^{1,\alpha}$

Let U be a domain, we recall the definition of $C^{1,\alpha}(\bar{U})$.

$$C^{0,\alpha}(\bar{U}) = \{f \in C(\bar{U}), \quad \sup_{\substack{x,y\in\bar{U}\\ x\neq y}} \frac{|f(x)-f(y)|}{|x-y|^\alpha} < +\infty\},$$

$$C^{m,\alpha}(\bar{U}) = \{f \in C^m(\bar{U}), \quad D^\beta f \in C^{0,\alpha}(\bar{U}) \ \ \forall\beta \text{ with } |\beta| = m\}.$$

A curve is $C^{1,\alpha}$ if locally it is the graph of a $C^{1,\alpha}$ functional.

The main result of this section is the

THEOREM 6.1 *Let (u, K) be a minimizer of the Mumford-Shah functional. Let P be a flat point of K. Assume that there is an open neighborhood ω of P such that ω intersects only one connected component of K. Then, there is a neighborhood of P in which K is a $C^{1,\alpha}$ arc for all $\alpha < 1/2$.*

Let $\phi_P : r \mapsto \int_{B_r(P)\setminus K} |\nabla u|^2$. With the same tools used for theorem 3.1, we can prove the weak monotonicity formula:

LEMMA 6.2 *Assume that for any $R < R_0$ there is a diameter $[A, B]$ of the ball $B_R(P)$ such that*

$$K \cap B_R(P) \subset \{X \in B_R(P), d(X, [A, B]) < lR\}$$

where l, $0 < l < 1$ is a constant. Assume that only one component of K intersects B_{R_0} and that K intersects both connected components of $\partial B_R(P) \cap \{X \in B_R(P), d(X, [A, B]) < lR\}$ then there is a $\beta = \beta(l)$ and a constant $C(R_0, \beta)$ such that

$$\phi_P(R) \leq CR^\beta.$$

The constant β is such that $\beta(l) \to 2$ as $l \to 0$.

An argument by contradiction allows us to prove that if $\int_{B_R} |\nabla u|^2$ is of order R^β then, the length of K in $B_R(P)$, $P \in K$ is that of a straight line passing through P up to R^β order terms:

LEMMA 6.3 *Under the hypotheses of lemma 6.2 and for* $l < \frac{\sqrt{3}}{2}$ *there is a constant* $C(R_0, \beta)$ *such that for* $R < R_0$*:*

$$\mathcal{H}^1(K \cap B_R(P)) \leq 2R + CR^\beta. \tag{16}$$

Proof of theorem 6.1.
Lemma 6.3 gives a strong constraint on K. If this is true, then K needs to be very flat in a neighborhood of P. The estimate in lemma 6.3 is derived for any $\beta < 2$ and is locally uniform on a Jordan arc. This implies the local $C^{1,\alpha}$ regularity for any $\alpha < 1/2$ [5].

7. Further regularity

In this section we will prove in theorem 7.2 that if K is $C^{1,\alpha}$ then it is $C^{1,1}$. We will see, then, that if K is $C^{1,\alpha}$ and if g is $C^{k,\mu}$ then K is $C^{k+2,\mu}$. In the following, (u, K) is a minimizer of the Mumford-Shah functional.

PROPOSITION 7.1 *Let* P *be a point of* K *such that in a neighborhood* $B_{4R}(P)$ *of* P *the edge set* K *is a* $C^{1,\alpha}$ *curve:* $\{(x,y),\ y = f(x),\ x \in I\}$. K *defines two sets* $B_R^+ = \{(x,y) \in B_R(P),\ y \geq f(x)\}$ *and* $B_R^- = \{(x,y) \in B_R(P),\ y \leq f(x)\}$. *The minimizer* u *can be extended on each side of* K *such that:*

$$u \in C^{1,\alpha}(B_R^+) \text{ and } u \in C^{1,\alpha}(B_R^-). \tag{17}$$

Proof. We will prove that u is $C^{1,\alpha}$ in $B_R^+ = \{(x,y) \in B_{2R}(P),\ y \geq f(x)\}$. The minimizer u satisfies $-\Delta u + u - g = 0$ in the interior of B_{4R}^+. In the sense of lemma 3.1 in [7] we can say that u satisfies Neumann boundary conditions $\frac{\partial u}{\partial \nu} = 0$ on K. We want to prove that a solution of $-\Delta u + u - g = 0$ in B_{4R}^+ with $\frac{\partial u}{\partial \nu} = 0$ on K is $C^{1,\alpha}$ in B_R^+. Since g is L^∞, the $C^{1,\alpha}$ regularity in the case of Dirichlet conditions is given by theorem 5.5.5′ of [16]. Following [16], it is possible in fact to prove the $C^{1,\alpha}$ regularity in the case of Neumann boundary conditions [5].

THEOREM 7.2 *Let* P *be a point of* K *such that in a neighborhood* U *of* P *the edge set* K *is a* $C^{1,\alpha}$ *curve, then* K *is* $C^{1,1}$ *in* U.

Proof. In some sense, it can be proved that the curvature of K is equal to the jump of $(u-g)^2 + |\nabla u|^2$ across K ([17] and [5]). Proposition 7.1 says that this is bounded and therefore K is $C^{1,1}$. Then a bootstrap argument with classical elliptic estimates gives theorem 1.4. Moreover, using section 5., theorems 7.2 and 6.1 give theorem 1.1.

REMARK 7.3 *Notice that it is easy to construct an example of an* L^∞ *function* g *such that* K *is* $C^{1,1}$ *and not* C^2.

8. Regularity almost everywhere

In this section we will not make any assumption on the connected components of K and we will prove theorem 1.3. Since K is a countably rectifiable set, for $\mathcal{H}^1$ almost every point P of K we have

a) there is a tangent line to K,

b) there is a one-to-one Lipschitz map $\psi : [0,1] \to \Omega$ such that $\psi([0,1]) \subset K$ and $P \in \psi((0,1))$.

In the following we will assume that P satisfies a) and b).

LEMMA 8.1 *The blow-up at P is a global-local minimizer of type (ii).*

Proof. Because of the existence of a tangent line L, the blow-up limit (u_0, K_0) is such that $K_0 \subset L$. Property b) implies that $L \subset K_0$. Then $K_0 = L$ and a straightforward argument implies that u_0 is constant on both sides of L.

LEMMA 8.2 *For any η there is a r_0 such that for all $r < r_0$, $\mathcal{H}^1(K \cap B_r(P)) \leq (2+\eta)r$.*

Proof. This is a consequence of (11).

LEMMA 8.3 *For all $r < r_0$, we have $|\{\rho,\ r > \rho > 0,\ \#(K \cap \partial B_r(P)) = 2\}| \geq r(1-\eta)$.*

Proof. Property (b) above implies that $\#(K \cap \partial B_r(P)) \geq 2$. Then lemma 8.2 gives the result.

Lemma 8.1 and lemma 8.3 allows to prove:

PROPOSITION 8.4 *For every $\beta < 2$, there is a constant C'' and a R_0 such that for $r < R < R_0$*

$$\frac{\phi_P(r)}{r^\beta} \leq \frac{\phi_P(R)}{R^\beta} + C'' R^{2-\beta}. \tag{18}$$

As in section 6. an estimation on $\frac{\phi_P(r)}{r}$ gives an information on K:

LEMMA 8.5 *There exists a function $m(\epsilon, r)$ such that*

$\lim_{(0,0)} m(\epsilon, r) = 0$ and:

if there is a straight line L passing through P and a $r > 0$ such that $K \cap B_r(P) \subset \{X,\ d(X,L) \leq r/4\}$ then

$$\frac{\phi_P(r)}{r} \leq \epsilon \Rightarrow \exists L' \text{ straight line}, K \cap B_r(P) \subset \{X,\ d(X,L') \leq m(\epsilon,r)r\}. \tag{19}$$

Proof. The proof comes from a contradiction argument. If this is not true then one can construct a blow-up sequence such that the blow-up limit satisfies $\phi \equiv 0$ and conditions on K which prevent it to be of type (i), (ii) or (iii), a contradiction.

We will prove now the:

Proposition 8.6 *For $\mathcal{H}^1$ almost every P in K there is a ball $B_\rho(P)$ which intersects exactly one connected component of K.*

Proof. For λ, ϵ and R small, proposition 8.4 and lemma 8.5 allow us to get by induction that for all $M \in \psi((0,1)) \cap B_{R/2}(P)$ we have

$$\frac{\phi_M(\lambda^k R)}{\lambda^k R} \leq \epsilon/2^k. \tag{20}$$

and that $\exists L_M^k$ such that

$$K \cap B_{\lambda^k R}(M) \subset \{X,\ d(X, L_M^k) \leq m(\epsilon/2^k, \lambda^k R)\lambda^k R\}.$$

The induction allows us to see that $K \cap B_{R/2}$ is a subset of the union of the balls $B_{\lambda^{k+1}R}(M)$ for $M \in \psi((0,1)) \cap B_{R/2}(P)$. In particular, this means that $K \cap B_{R/2}(P) \subset \{X,\ d(X, \psi((0,1)) \leq \lambda^{k+1}R\}$. Therefore, letting k go to $+\infty$ we deduce that

$$K \cap B_{R/2}(P) \subset \psi((0,1)). \tag{21}$$

More precisely we have $K \cap B_{R/2}(P) = \psi((0,1)) \cap B_{R/2}(P)$ which completes the proof of the proposition.

Theorem 1.3 is then a consequence of theorem 6.1 and 7.2.

Acknowledgment. I am very grateful to J.-M. Morel for many discussions and his interest in this work. I would like to thank him, F. Dibos and E. Séré for introducing me to this problem.

References

[1] H. W. Alt, L. A. Caffarelli & A. Friedman, "Variational Problems With Two Phases and Their Free Boundary", Trans. Am. Math. Soc. 282, 431–461 (1984).

[2] L. Ambrosio, "Existence theory for a new class of variational problems", Arch. Rat. Mech. Anal. **111** (1990), 291–322.

[3] L. Ambrosio & D. Pallara, "Partial regularity of Free discontinuity sets I", to appear.

[4] L. Ambrosio, N. Fusco & D. Pallara, "Partial regularity of Free discontinuity sets II", to appear.

[5] A. Bonnet, "On the regularity of edges in image segmentation", to appear.

[6] G. Congedo & I. Tamanini, "On the existence of solutions to a problem in multidimensional segmentation", Ann. Inst. Henri Poincaré, vol. 8, 2, 1991, P. 175–195.

[7] G. Dal Maso, J.-M. Morel & S. Solimini. "A variational method in image segmentation: existence and approximation results", Acta Mathematica, Vol. 168, 89–151, 1992.

[8] G. David & S. Semmes, "On the singular sets of minimizers of the Mumford-Shah functional", to appear in J. Math. Pures & Appl.

[9] G. David, "C^1-arcs for minimizers of the Mumford-Shah functional", to appear.

[10] F. Dibos, "Uniform rectifiability of image segmentations obtained by a variational method", J. Math. Pures & Appl. 73 (1994), 389–412.

[11] E. De Giorgi, M. Carriero & A. Leaci "Existence theorem for a minimum problem with free discontinuity set", Arch. Rat. Mech. Anal. **108** (1989), 195–218.

[12] H. Federer, "Geometric measure theory, Springer-Verlag", 1969.

[13] G. Hardy, JE. Littlewood & G. Pólya, "Inequalities", Second Edition, Cambridge University Press.
[14] U. Massari & I. Tamanini, "Regularity properties of optimal segmentations", Journ. reine angew. Math. 420 (1991), 61–84.
[15] J.-M. Morel & S. Solimini, "Variational methods in Image Segmentation".
[16] C. B. Morrey Jr. "Multiple Integrals in the Calculus of Variations", Springer-Verlag, 1966.
[17] D. Mumford & J. Shah, "Optimal Approximations by Piecewise Smooth Functions and Associated Variational Problems", Comm. on Pure and Appl. Math. Vol XLII no. 4, 1989.
[18] L. Simon, "Lectures on Measure Theory", Proceedings of the Centre for Mathematical Analysis, Australian National University, Vol 3, 1983.

Progress in Nonlinear Differential Equations
and Their Applications, Vol. 25

Capacity and Dirichlet Problems in varying Domains

GIANNI DAL MASO

SISSA, via Beirut 4, 34013 Trieste, Italy

ABSTRACT. The asymptotic behaviour of solutions of non-linear second order elliptic equations with Dirichlet boundary conditions in perforated domains is studied by means of a suitable notion of non-linear capacity.

Let Ω be a bounded open set in $\mathbf{R}^n$, and let $A: H_0^{1,p}(\Omega) \to H^{-1,q}(\Omega)$ be a monotone operator of the form

$$Au = -\operatorname{div}\big(a(x, Du)\big)\,,$$

with $2 \leq p \leq n$, $1/p + 1/q = 1$. The case $1 < p < 2$ can be treated by similar arguments with minor changes in the hypotheses. The case $p > n$ is less interesting, because of the compact embedding of $H_0^{1,p}(\Omega)$ into $C^0(\overline{\Omega})$. We assume that there exist two constants $c_0 > 0$ and $c_1 > 0$ such that

$$\big(a(x,\xi_1) - a(x,\xi_2), \xi_1 - \xi_2\big) \geq c_0|\xi_1 - \xi_2|^p\,, \tag{1}$$

$$|a(x,\xi_1) - a(x,\xi_2)| \leq c_1\big(1 + |\xi_1| + |\xi_2|\big)^{p-2}|\xi_1 - \xi_2| \tag{2}$$

for every $x \in \Omega$ and for every $\xi_1, \xi_2 \in \mathbf{R}^n$, where $(\cdot,\cdot)$ denotes the scalar product in $\mathbf{R}^n$. Without loss of generality we may assume also that

$$a(x,0) = 0\,. \tag{3}$$

Given a sequence (Ω_j) of open sets contained in Ω, we consider for every $f \in H^{-1,q}(\Omega)$ the sequence (u_j) of the solutions of the non-linear Dirichlet problems

$$\begin{cases} Au_j = f & \text{in } \Omega_j, \\ u_j \in H_0^{1,p}(\Omega_j). \end{cases} \tag{4}$$

By (1), (2), and (3) for every j problem (4) has one and only one solution u_j defined in Ω_j. We extend u_j to Ω by setting $u_j = 0$ on $\Omega \setminus \Omega_j$, and we consider (u_j) as a sequence in $H_0^{1,p}(\Omega)$. The problem is to describe the asymptotic behaviour of the sequence (u_j) as j tends to infinity.

This problem has been studied by several authors under suitable geometric conditions on the sets Ω_j. We refer to [9] and [13] for a wide bibliography on this subject. In the present paper we shall consider only those results that hold without any assumption on the sets Ω_j.

By using (1), (2), and a diagonal argument, it is easy to see that for every sequence (Ω_j) there exists a subsequence, still denoted by (Ω_j), such that for every $f \in H^{-1,q}(\Omega)$ the sequence (u_j) converges weakly in $H_0^{1,p}(\Omega)$ to some function u.

The following theorem characterizes u as the solution of a suitable boundary value problem. We recall that the *p-capacity* of a set $E \subseteq \Omega$ is defined as the infimum of $\int_\Omega |Du|^p dx$ over the set of all functions $u \in H_0^{1,p}(\Omega)$ such that $u \geq 1$ a.e. in a neighbourhood of E.

THEOREM 1 *Assume that for every $f \in H^{-1,q}(\Omega)$ the sequence (u_j) converges weakly in $H_0^{1,p}(\Omega)$ to some function u. Then there exist a non-negative measure μ on Ω, vanishing on all sets of p-capacity zero, and a Carathéodory function $b(x,s)$, increasing with respect to s, such that for every $f \in H^{-1,q}(\Omega)$ the limit function u is the solution of the problem*

$$\begin{cases} u \in H_0^{1,p}(\Omega) \cap L^p_\mu(\Omega)\,, \\ \langle Au, v\rangle + \displaystyle\int_\Omega b(x,u)v\,d\mu = \langle f, v\rangle \qquad \forall v \in H_0^{1,p}(\Omega) \cap L^p_\mu(\Omega)\,. \end{cases} \tag{5}$$

When A is linear and symmetric this result was proved in [2], [1], [12], [5], and [17] by using Γ-convergence techniques. In this case the function b has the special form $b(x,s) = a(x)s$. The same technique was used later in [8] to attack a non-linear case, where $A = \partial\Psi$ is the differential of a convex functional of the form $\Psi(u) = \int_\Omega \psi(x, Du)\,dx$, which is assumed to be even and homogeneous of degree p. In this case the the function b has the form $b(x,s) = a(x)|s|^{p-2}s$.

When A is linear and non-symmetric the problem is no longer equivalent to a minimum problem, and the proof of Theorem 1, with $b(x,s) = a(x)s$, requires a totally different technique (see [9]). When A is odd and homogeneous of degree $p-1$ the function b has the form $b(x,s) = a(x)|s|^{p-2}s$ (see [13] and [14])

The general case of a monotone operator satisfying only conditions (1), (2), and (3) was proved recently in [6]. In this case the function b is increasing and locally Hölder continuous with respect to s, and satisfies the estimates

$$s\,b(x,s) \geq c_2|s|^p\,, \qquad |b(x,s)| \leq c_3|s|^{p-1}\,, \tag{6}$$

where c_2 and c_3 are positive constants depending only on c_0 and c_1.

A very special situation covered by Theorem 1 is when $\Omega_j = \Omega_0$ for every j. Then u_j is equal to the solution u_0 of the problem

$$\begin{cases} Au_0 = f \qquad \text{in } \Omega_0\,, \\ u_0 \in H_0^{1,p}(\Omega_0)\,. \end{cases} \tag{7}$$

It is clear that in this case the limit function u coincides with u_0. In order to show that this does not contradict the fact that u must be also the solution of a problem of the form (5), we introduce the measure μ_0 defined by

$$\mu_0(E) = \begin{cases} 0, & \text{if} \quad C_p(E \setminus \Omega_0) = 0, \\ +\infty, & \text{if} \quad C_p(E \setminus \Omega_0) > 0, \end{cases}$$

where C_p denotes the p-capacity. It is possible to prove that u is the solution of (5), with $\mu = \mu_0$ and $b(x,s) = |s|^{p-2}s$, if and only if u coincides with the solution u_0 of (7) in Ω_0, and $u = 0$ in $\Omega \setminus \Omega_0$. This shows that the family of problems of the form (5) includes the classical Dirichlet problems of the form (7) as well as their limits.

It is possible to prove that the family of problems of the form (5) is stable with respect to the convergence of the solutions, as shown by the following theorem.

THEOREM 2 *Let* (μ_j) *be a sequence of non-negative measures on* Ω *vanishing on all sets of* p*-capacity zero, and let* (b_j) *be a sequence of Carathéodory functions satisfying* (6) *with constants* c_2 *and* c_3 *independent of* j, *and some other technical estimates, for which we refer to [6]. Then there exist an increasing sequence of integers* (j_k), *a non-negative measure* μ_∞ *on* Ω, *vanishing on all sets of* p*-capacity zero, and a Carathéodory function* b_∞ *satisfying* (6), *such that for every* $f \in H^{-1,q}(\Omega)$ *the sequence* (u_k) *of the solutions of* (5), *with* $\mu = \mu_{j_k}$ *and* $b = b_{j_k}$, *converges weakly in* $H_0^{1,p}(\Omega)$ *to the solution* u_∞ *of* (5), *with* $\mu = \mu_\infty$ *and* $b = b_\infty$.

The measure μ and the function b of Theorem 1 do not depend on f. They depend only on the sequence of sets (Ω_j). It turns out that the properties of (Ω_j) which are relevant in the construction of μ and b are related to the notion of A-capacity. If K is a compact set contained in Ω and s is a real number, the A*-capacity* of K in Ω relative to the constant s is defined as

$$C_A(K,s) = \int_{\Omega\setminus K} \big(a(x,Du),Du\big)\,dx\,,$$

where u, the A*-potential* of K in Ω relative to the constant s, is the weak solution of the Dirichlet problem

$$\begin{cases} Au = 0 & \text{in } \Omega\setminus K\,,\\ u = s \text{ in } \partial K\,, & u = 0 \text{ in } \partial\Omega\,. \end{cases}$$

If $Au = -\mathrm{div}\big(|Du|^{p-2}Du\big)$ is the p-Laplacian, then we have

$$C_A(K,s) = |s|^p C_p(K)\,,$$

where $C_p(K)$ is the p-capacity of the set K. In particular, when $p = 2$ and $Au = -\Delta u$ is the Laplacian, then $C_A(K,1)$ is the usual Newtonian capacity of the compact set K with respect to Ω.

It is possible to prove that C_A is an incresing set function (see [15]), i.e.,

$$K_1 \subseteq K_2 \qquad \Longrightarrow \qquad C_A(K_1,s) \le C_A(K_2,s)\,.$$

The idea of the proof is very simple when a, ∂K_1, ∂K_2, and $\partial\Omega$ are smooth enough. In this case we have

$$\begin{aligned} C_A(K_1,s) &= \int_{\Omega\setminus K_1} \big(a(x,Du_1),Du_1\big)\,dx = -\int_{\Omega\setminus K_1} \big(a(x,Du_1),D(s-u_1)\big)\,dx \\ &= -\int_{\partial(\Omega\setminus K_1)} \big(a(x,Du_1),(s-u_1)n\big)\,d\sigma = -s\int_{\partial\Omega} \big(a(x,\frac{\partial u_1}{\partial n}n),n\big)\,d\sigma\,, \end{aligned}$$

where u_1 is the A-potential of K_1 in Ω relative to the constant s and n is the outer unit normal. Similarly we prove that

$$C_A(K_2,s) = -s\int_{\partial\Omega} \big(a(x,\frac{\partial u_2}{\partial n}n),n\big)\,d\sigma\,,$$

where u_2 is the A-potential of K_2 in Ω relative to the constant s. Then

$$C_A(K_1, s) - C_A(K_2, s) = -s \int_{\partial\Omega} \Big(a(x, \frac{\partial u_1}{\partial n} n) - a(x, \frac{\partial u_2}{\partial n} n), n\Big)\, d\sigma\,. \tag{8}$$

By monotonicity we have

$$\Big(a(x, \frac{\partial u_1}{\partial n} n) - a(x, \frac{\partial u_2}{\partial n} n), \big(\frac{\partial u_1}{\partial n} - \frac{\partial u_2}{\partial n}\big) n\Big) \geq 0\,.$$

When $s > 0$, by the maximum principle it is easy to see that $u_1 \leq u_2$ in Ω, hence $\frac{\partial u_1}{\partial n} \geq \frac{\partial u_2}{\partial n}$ in $\partial\Omega$. This implies that

$$\Big(a(x, \frac{\partial u_1}{\partial n} n) - a(x, \frac{\partial u_2}{\partial n} n), n\Big) \geq 0\,,$$

so that (8) gives $C_A(K_1, s) - C_A(K_2, s) \leq 0$. The same conclusion holds if $s < 0$. For the complete proof in the non-smooth case we refer to [15].

The notion of A-capacity can be extended to all open subsets U of Ω by

$$C_A(U, s) = \sup\{C_A(K, s) : K \text{ compact, } K \subseteq U\}\,,$$

and to arbitrary subsets E of Ω by

$$C_A(E, s) = \inf\{C_A(U, s) : U \text{ open, } U \supseteq E\}\,.$$

It is possible to prove that C_A is countably subadditive (see [15]), i.e.,

$$E \subseteq \bigcup_{j=1}^{\infty} E_j \qquad \Longrightarrow \qquad C_A(E, s) \leq \sum_{j=1}^{\infty} C_A(E_j, s)\,.$$

Moreover, it is possible to prove that C_A is a Choquet capacity (see [15]), i.e., $C_A(\cdot, s)$ is continuous along all increasing sequences of subsets of Ω and along all decreasing sequences of compact subsets of Ω.

Assume now that the sequence (Ω_j) of open sets satisfies the following property: there exists a function $\alpha(K, s)$ such that

$$\lim_{j\to\infty} C_A(K \setminus \Omega_j, s) = \alpha(K, s) \tag{9}$$

for every $s \in \mathbf{R}$ and for "almost every" compact set K contained in Ω. This is a very weak assumption, since it is possible to prove that every sequence (Ω_j) contains a subsequence which satisfies (9). The function $\alpha(\cdot, s)$ can be interpreted as an asymptotic capacity relative to the sequence (Ω_j) (see [3]).

Assume, in addition, that there exists a bounded measure λ on Ω, vanishing on all sets of zero p-capacity, such that

$$\alpha(Q, s) \leq k(s)\lambda(Q) \tag{10}$$

for every $s \in \mathbf{R}$ and for every cube Q in Ω, where $k(s)$ is a constant depending on s. Let $Q_\rho(x)$ be the closed cube with centre x and side 2ρ. Then the following theorem holds.

THEOREM 3 *Assume that* (Ω_j) *satisfies* (9) *and* (10). *Then the limit*

$$\lim_{\rho\to 0} \frac{\alpha(Q_\rho(x), s)}{\lambda(Q_\rho(x))} = g(x, s)$$

exists for λ*-a.e.* $x \in \Omega$ *and for every* $s \in \mathbf{R}$. *Moreover, for every* $f \in H^{-1,q}(\Omega)$ *the solutions* u_j *of problems* (4) *converge weakly in* $H_0^{1,p}(\Omega)$ *to the solution* u *of problem* (5) *with*

$$\mu(E) = \textstyle\int_E g(x,-1)\, d\lambda(x) \qquad \text{and} \qquad b(x,s) = \frac{1}{s}\frac{g(x,-s)}{g(x,-1)}.$$

When A is linear and symmetric this result was proved in [4]. The case $A = \partial\Psi$, with Ψ convex, even, and p-homogeneous, was proved in [8]. When A is linear and non-symmetric the proof can be found in [10]. The case of a general monotone operator satisfying only (1), (2), and (3) is studied in [18], when λ is the Lebesgue measure, and in [16] for a wide class of measures λ. The general case of an arbitrary bounded measure λ is considered in [11].

If condition (9) is fulfilled, but condition (10) is not satisfied, it is still possible to construct the measure μ and the function b by a capacitary method. Let β be the regularized version of α defined for every open set U in Ω by

$$\beta(U,s) = \sup\{\alpha(K,s) : K \text{ compact, } K \subseteq U\},$$

and for every subset E of Ω by

$$\beta(E,s) = \inf\{\beta(U,s) : U \text{ open, } U \supseteq E\}.$$

Let us define for every Borel set $E \subseteq \Omega$ and for every $s \in \mathbf{R}$

$$\nu(E,s) = \sup \sum_{i\in I} \beta(E_i, s), \tag{11}$$

where the supremum is over all finite Borel partitions $(E_i)_{i\in I}$ of E. Then for every $s \in \mathbf{R}$ the set function $\nu(\cdot, s)$ is the smallest Borel measure on Ω such that $\nu(E,s) \geq \beta(E,s)$ for every Borel set $E \subseteq \Omega$. Let λ be the measure defined by

$$\lambda(E) = \nu(E,1) \tag{12}$$

for every Borel set $E \subseteq \Omega$. It is easy to prove that

$$c_4|s|^p\lambda(E) \leq \nu(E,s) \leq c_5|s|(1+|s|^{p-1})\lambda(E), \tag{13}$$

where c_4 and c_5 are constants depending only on c_0, c_1, p, and Ω. Then the following theorem holds.

THEOREM 4 *Assume that* (Ω_j) *satisfies* (9), *and let* $\nu(E,s)$ *and* $\lambda(E)$ *be the measures defined by* (11) *and* (12). *Then* (13) *implies that there exists a Carathéodory function* $g(x,s)$ *such that*

$$\nu(E,s) = \int_E g(x,s)\, d\lambda(x)$$

for every Borel set $E \subseteq \Omega$ *and for every* $s \in \mathbf{R}$. *Moreover, for every* $f \in H^{-1,q}(\Omega)$ *the solutions* u_j *of problems* (4) *converge weakly in* $H_0^{1,p}(\Omega)$ *to the solution* u *of problem* (5) *with* $\mu = \lambda$ *and* $b(x,s) = \frac{1}{s}g(x,-s)$.

When A is linear and symmetric this result was proved in [7]. The case $A = \partial\Psi$, with Ψ convex, even, and p-homogeneous, was proved in [8]. When A is linear and non-symmetric the proof can be found in [10]. The case of a general monotone operator satisfying only (1), (2), and (3) is studied in [11].

References

[1] ATTOUCH H.: Variational Convergence for Functions and Operators. Pitman, London, 1984.

[2] ATTOUCH H., PICARD C.: Variational inequalities with varying obstacles: the general form of the limit problem. *J. Funct. Anal.* **50** (1983), 329–386.

[3] BAXTER J.R., JAIN N.C.: Asymptotic capacities for finely divided bodies and stopped diffusions. *Illinois J. Math.* **31** (1987), 469–495.

[4] BUTTAZZO G., DAL MASO G., MOSCO U.: A derivation theorem for capacities with respect to a Radon measure. *J. Funct. Anal.* **71** (1987), 263–278.

[5] BUTTAZZO G., DAL MASO G., MOSCO U.: Asymptotic behaviour for Dirichlet problems in domains bounded by thin layers. *Partial Differential Equations and the Calculus of Variations, Essays in Honor of Ennio De Giorgi*, 193–249, *Birkhäuser, Boston*, 1989.

[6] CASADO DIAZ J., GARRONI A.: Asymptotic behaviour of nonlinear elliptic systems on varying domains. In preparation.

[7] DAL MASO G.: Γ-convergence and μ-capacities. *Ann. Scuola Norm. Sup. Pisa Cl. Sci. (4)* **14** (1987), 423–464.

[8] DAL MASO G., DEFRANCESCHI A.: Limits of nonlinear Dirichlet problems in varying domains. *Manuscripta Math.* **61** (1988), 251–278.

[9] DAL MASO G., GARRONI A.: New results on the asymptotic behaviour of Dirichlet problems in perforated domains. *Math. Mod. Meth. Appl. Sci.* **3** (1994), 373–407.

[10] DAL MASO G., GARRONI A.: The capacity method for asymptotic Dirichlet problems. Preprint SISSA, Trieste, 1994.

[11] DAL MASO G., GARRONI A., SKRYPNIK I.V.: A capacity method for the asymptotic analysis of Dirichlet problems for monotone operators. Preprint SISSA, Trieste, 1996.

[12] DAL MASO G., MOSCO U.: Wiener's criterion and Γ-convergence. *Appl. Math. Optim.* **15** (1987), 15–63.

[13] DAL MASO G., MURAT F.: Dirichlet problems in perforated domains for homogeneous monotone operators on H^1_0. *Calculus of Variations, Homogenization and Continuum Mechanics (CIRM-Luminy, Marseille, 1993)*, 177–202, *World Scientific, Singapore*, 1994.

[14] DAL MASO G., MURAT F.: Asymptotic behaviour and correctors for Dirichlet problems in perforated domains with homogeneous monotone operators. Preprint SISSA, Trieste, 1996.

[15] DAL MASO G., SKRYPNIK I.V.: Capacity theory for monotone operators. *Potential Anal.*, to appear.

[16] DAL MASO G., SKRYPNIK I.V.: Asymptotic behaviour of nonlinear Dirichlet problems in perforated domains. *Ann. Mat. Pura Appl.*, to appear.

[17] NOTARANTONIO L.: Asymptotic behaviour of Dirichlet problems on a Riemannian manifold. *Ricerche Mat.* **41** (1992), 327–367.

[18] SKRYPNIK I.V.: Homogenization of nonlinear Dirichlet problems in perforated domains of general structure. *Mat. Sb. (N.S.)*, to appear.

Progress in Nonlinear Differential Equations
and Their Applications, Vol. 25

General Growth Conditions and Regularity

PAOLO MARCELLINI

Dipartimento di Matematica "U. Dini", Università di Firenze
Viale Morgagni 67/A, 50134 Firenze, Italy

1. Introduction

We study the local regularity of minimizers of integral functionals of the calculus of variations of the type

$$F(v) = \int_{\Omega} f(Dv)dx, \tag{1}$$

where Ω is an open set of $\mathbb{R}^n$ $(n \geq 2)$, Du is the $N \times n$ matrix of the gradient of $u : \Omega \to \mathbb{R}^N$ and $f : \mathbb{R}^{N \times n} \to \mathbb{R}$ is a given convex function.

Since we will assume that f has at least quadratic growth at infinity (but not necessarily polynomial growth), by a *minimizer* of the integral functional (1) we mean a function $u \in W^{1,2}_{\text{loc}}(\Omega; \mathbb{R}^N)$ such that $f(Du) \in L^1_{\text{loc}}(\Omega)$, with the property that $F(u) \leq F(u + \varphi)$ for every $\varphi \in C^1_0(\Omega; \mathbb{R}^N)$.

In recent years interest has increased in the study of the regularity of solutions of elliptic equations and systems and of minimizers of integral functionals of the calculus of variations under *general* and *non-standard growth conditions.* Let us consider here a list of such integrals; we begin with an example of an integral with *anisotropic growth*:

$$F(v) = \int_{\Omega} \left\{ \left(1 + |Dv|^2\right)^{p/2} + \sum_{i,\alpha} |v^{\alpha}_{x_i}|^{q^{\alpha}_i} \right\} dx, \tag{2}$$

for some exponents $q^{\alpha}_i > p \geq 2$, for $i = 1, 2, \ldots, n$, $\alpha = 1, 2, \ldots, N$.

The following functional

$$F(v) = \int_{\Omega} \left(1 + |Dv|^2\right)^{\alpha(x)/2} dx, \text{ with } 2 \leq p \leq \alpha(x) < q, \tag{3}$$

is an example of a functional which satisfies the *p, q-growth condition.* Also the integral in (2) has a growth of type p, q, with p and q respectively equal to the minimum and the maximum of $\{p, q^{\alpha}_i : i = 1, 2, \ldots, n, \ \alpha = 1, 2, \ldots, N\}$. An integral of the type

$$\int_{\Omega} f(Dv)dx, \text{ with } f(\xi) \sim |\xi|^p \log(1 + |\xi|) \text{ as } |\xi| \to +\infty \tag{4}$$

satisfies the *p, q-growth* condition with q arbitrarily close to p. When we have $q = p$, then we use the classical terminology of *natural growth conditions.*

If the growth with respect to the gradient Dv is not bounded by a power and, for example, is an *exponential growth* of the type

$$\int_{\Omega} f(Dv)dx, \text{ with } f(\xi) \sim \exp\left(|\xi|^{\alpha}\right) \text{ as } |\xi| \to +\infty \tag{5}$$

for some $\alpha > 0$, then we have a problem with *general growth conditions.*

A reason to classify elliptic problems with respect to growth conditions is that growth conditions involving powers give rise to *uniformly* elliptic problems (as $|\xi| \to +\infty$); on the other hand more general growth conditions, as for example exponential growth conditions like in (5), give rise to *non-uniformly* elliptic problems. Note that integrands with exponential growth does not satisfy the Δ_2 condition ($f(2\xi) \leq \text{const} \cdot f(\xi)$, as $|\xi| \to +\infty$; see [25]).

The regularity theory for elliptic equations and for integrals of the calculus of variations with non-standard growth conditions has been first considered and studied by Marcellini in [35], [36]. Examples of singular weak minimizers have been given by Giaquinta and Marcellini [15], [34] (see also [22], [36]). Roughly speaking, under *p, q-growth* conditions we have regularity (at least in the scalar case with $q \geq p \geq 2$) if the ratio q/p is sufficiently close to 1 in dependence of the dimension n, while we may have discontinuities if q/p is large. A sharp condition has been introduced by Boccardo, Marcellini and Sbordone [5] to study the special anisotropic case of the type in (2) in the scalar case; further results for anisotropic functionals and equations have been given in [4], [13], [14], [22], [24], [35], [48] and in [1], where partial regularity for elliptic anisotropic systems has been considered. Many other papers deal with non-standard growth conditions; for example: [2], [3], [7], [12], [28], [29], [31], [43], [49], [50]. In [37] has been obtained, in the scalar case, the local Lipschitz-continuity and the $C^{1,\alpha}_{\text{loc}}$ regularity of minimizers of integral-functionals with exponential "*slow*" growth; precisely, regularity of minimizers of integrals of the type in (5), of growth $\exp|\xi|^{\alpha}$, by assuming that $\alpha < 2\log[n/(n-2)] / \log n$ (no restrictions on α if $n = 2$). We mention also the Lipschitz and the $C^{1,\alpha}_{\text{loc}}$ regularity obtained for systems in [6] (quadratic growth), [17], [45], [51] (*p-growth*, with $p \geq 2$). Specific for functions depending on the modulus of the gradient, i.e. $f(\xi) = g(|\xi|)$, we mention the regularity results in [7] (with functions g of class Δ_2); the local boundedness and the Hölder continuity of minimizers, again with g of class Δ_2, studied in the scalar case in [12], [40], [41], [43]; the everywhere regularity for a class of vector-valued variational problems, studied in [38] (see also Section 3 in this paper).

Regularity of weak solutions under *natural growth conditions* has an older story; main references are the books [16], [26]. More recent Lipschitz-continuity results can be found in [6], [9], [11], [33], [52]. The uniformly elliptic case has been considered in Section 4 of [47] and more recently in [31]. Gradient estimates for classical solutions of non-uniformly elliptic equations have been given in [23], [27], [46], [47], with particular interest in the mean curvature equation and in its generalizations.

In this paper we present two new regularity results, one in the scalar case and the second in the vector-valued case, specific for problems with general growth conditions. These results can be applied not only to integrals of the type in (5), of growth $\exp|\xi|^\alpha$ with α positive and close to zero, but also, for example, to integrands of "*large*" growth, of the type

$$f(\xi) = \exp(|\xi|^p), \text{ with } p \geq 2, \tag{6}$$

or even to any finite composition of functions of the type

$$f(\xi) = (\exp(\ldots(\exp(\exp|\xi|^2)^{p_1})^{p_2})\ldots)^{p_k}, \text{ with } p_i \geq 1, \forall i = 1, 2, \ldots, k. \tag{7}$$

2. The scalar case

Here we assume that $f : \mathbb{R}^n \to \mathbb{R}$ is a non-negative function of class $C^2(\mathbb{R}^n)$ satisfying the following *general growth conditions*: there exist two increasing functions $g_1, g_2 : [0, +\infty) \to [0, +\infty)$ (not identically zero) and a constant $C > 0$ such that

$$g_1(|\xi|)|\lambda|^2 \leq \sum_{i,j} f_{\xi_i\xi_j}(\xi)\lambda_i\lambda_j \leq g_2(|\xi|)|\lambda|^2, \forall \lambda, \xi \in \mathbb{R}^n, \tag{8}$$

$$g_2(t) \cdot t^2 \leq C\left\{1 + \int_0^t \sqrt{g_1(s)}\, ds\right\}^\alpha, \forall t \geq 0, \tag{9}$$

$$g_2(|\xi|) \cdot |\xi|^2 \leq C\{1 + f(\xi)\}^\beta, \forall \xi \in \mathbb{R}^n, \tag{10}$$

for some real exponents α, β such that

$$2 \leq \alpha < \frac{2n}{n-2}, \qquad 1 \leq \beta < \frac{2}{n} \cdot \frac{\alpha}{\alpha - 2} \tag{11}$$

(of course, when $n = 2$, in (11) α can be any real number greater than or equal to 2 and, if $\alpha = 2$, then β can be an arbitrary number greater than or equal to 1).

Let us denote by B_ρ, B_R balls of radii respectively ρ and R ($\rho < R$) contained in Ω and with the same center. The following theorem, whose proof will appear in [39], holds:

THEOREM A *Under the general growth conditions* (8), (9), (10) *and* (11), *every minimizer of the integral functional* (1) *is locally Lipschitz-continuous and, for every* $R > \rho > 0$, *there exists a constant* $c = c(n, \rho, R)$ *such that*

$$\|Du\|^2_{L^\infty(B_\rho;\mathbb{R}^n)} \leq c \left\{\int_{B_R} [1 + f(Du)]\, dx\right\}^\theta \tag{12}$$

for some exponent $\theta = \theta(\alpha, \beta, n) \geq 1$.

Once we have the estimate for the L^∞ – *norm* of the gradient, then the behavior as $t \to +\infty$ of $g_1(t)$, $g_2(t)$ becomes irrelevant to obtain further regularity of solutions. Thus, for example, we obtain the following regularity result in the classical sense:

COROLLARY B *Let us assume that the general growth conditions* (8), (9), (10) *and* (11) *hold with strictly positive functions* $g_1, g_2 : [0,+\infty) \to (0,+\infty)$. *Then, if* $f \in C^{k,\alpha}_{\text{loc}}(\mathbb{R}^n)$ *for some* $k \geq 2$, *every minimizer of the integral* (1) *is of class* $C^{k,\alpha}_{\text{loc}}(\Omega)$.

3. Everywhere regularity in the convex vectorial case

In this section we are interested in the local regularity (the local Lipschitz-continuity and then, as a consequence, the $C^{1,\alpha}_{\text{loc}}$ and C^{∞} regularity) of weak solutions of elliptic equations and systems of the type

$$\sum_i \frac{\partial}{\partial x_i} a_i^{\alpha}(Du) = 0 , \forall \alpha = 1,2,\ldots,N, \tag{13}$$

in an open set Ω of $\mathbb{R}^n$ ($n \geq 2$), where Du is the gradient of a vector-valued function $u : \Omega \to \mathbb{R}^N$ ($N \geq 1$). We assume that the vector field $a_i^{\alpha} : \mathbb{R}^{nN} \to \mathbb{R}^{nN}$ is the gradient of a function $f : \mathbb{R}^{nN} \to \mathbb{R}$. Under our convexity conditions, every weak solution of the elliptic system (13) is a minimizer of the integral

$$F(v) = \int_{\Omega} f(Dv)dx \tag{14}$$

and vice-versa. Recall that by a em weak solution to the elliptic system (13) we mean a function $u \in W^{1,2}_{\text{loc}}(\Omega,\mathbb{R}^N)$ such that $f(Du) \in L^1_{\text{loc}}(\Omega)$, which satisfies (13) in the sense of distributions. Similarly, as we said in the introduction, by a *minimizer* of the integral (14) we mean a function $u \in W^{1,2}_{\text{loc}}(\Omega,\mathbb{R}^N)$, such that $f(Du) \in L^1_{\text{loc}}(\Omega)$, with the property that $F(u) \leq F(u+\varphi)$ for every $\varphi \in C^1_0(\Omega,\mathbb{R}^N)$.

In this context of vector-valued problems *in general* it is natural to look for *almost everywhere regularity* of minimizers, introduced in the late 60's (precisely in 1968–1969) by Morrey [42], Giusti-Miranda [21] and Giusti [18]. In fact, counterexamples to the *everywhere regularity* of weak solutions of elliptic systems have been given, first by De Giorgi [8] in 1968, then in the same year by Giusti-Miranda [20] and, more specifically for minimizers of integrals of the type of (14) with f of quadratic growth, by Nečas [44] (see also Example 3.4 of Chapter II in [16]) in 1975–1977.

Nevertheless Uhlenbeck [52], in a well-known paper in 1977, proved that, for every fixed $p \geq 2$, minimizers of the integral

$$\int_{\Omega} | Dv |^p dx \tag{15}$$

are *everywhere* smooth, precisely, they are of class $C^{1,\alpha}_{\text{loc}}(\Omega,\mathbb{R}^N)$.

Therefore, the study of the everywhere regularity of vector-valued problems explains our restriction to integral functionals of the type

$$\int_{\Omega} f(Dv)dx, \text{ with } f(Dv) = g(| Dv |) \tag{16}$$

(here we pose $f(\xi) = g(|\xi|)$ for every $\xi \in \mathbb{R}^{N\times n}$; with $\xi = (\xi_i^\alpha)$, $i = 1, 2, \ldots, n$, $\alpha = 1, 2, \ldots, N$), where $g : [0, +\infty) \to \mathbb{R}$ is a convex function.

Similarly, in terms of systems, we restrict ourselves to elliptic systems of the form

$$\sum_i \frac{\partial}{\partial x_i} a_i^\alpha(Du) = 0 \, , \; \forall \, \alpha = 1, 2, \ldots, N, \text{ with } a_i^\alpha(Du) = a(|Du|) u_{x_i}^\alpha \, , \tag{17}$$

with $a : [0, +\infty) \to [0, +\infty)$ related to g in (16) by the condition

$$a(t) = g'(t)/t \, . \tag{18}$$

As we already pointed out in Section 1, in recent years interest has increased in the study of regularity of solutions of elliptic equations and systems under *general growth conditions*. In particular, being more related to the context considered in this section, we mention the recent regularity results for the specific integral functional ((16) with $f(\xi) = \exp(|\xi|^2)$, given by Lieberman [32] in the scalar case $N = 1$, and by Duc-Eells [10] in the vector-valued case $N \geq 1$.

The main assumption of this section is the following *nonoscillatory condition:*

$$\begin{cases} \textit{(i) the function } a : [0, +\infty) \to [0, +\infty) \textit{ is increasing;} \\ \textit{(ii) for every } \alpha > 1 \textit{ the limit } \lim\limits_{t \to +\infty} \frac{a'(t) \cdot t}{[a(t)]^\alpha} \textit{ exists.} \end{cases} \tag{19}$$

To avoid trivial situations, we consider functions $a(t)$ not identically equal to zero; however, we do not assume that $a(0) > 0$ and, more generally, we allow $a(t)$ to be equal to zero in $[0, t_0]$, with $t_0 > 0$.

Of course ((19) implies a type of "growth" condition: first (i) implies that $a(t) \geq a(1)$ for all $t \geq 1$ (so that f in (16) has at least quadratic growth, like in the context (15), with $p \geq 2$, considered by Uhlenbeck [52] ; secondly, the derivative $a'(t)$ can be bounded in terms of the *α-power* of $a(t)$. However these growth restrictions are weak enough to be satisfied, for example, not only by the family of functions of exponential growth in (5) with $\alpha > 0$ small, but also by the integrands f in (6), (7).

Theorem C and Corollary D that follow have been proved in [38].

THEOREM C *Under the nonoscillatory condition* (19), *let u be a weak solution to the elliptic system* (13) *(or a minimizer to the integral functional* (14)*). Then $u \in W^{1,\infty}_{\text{loc}}(\Omega; \mathbb{R}^N)$ and, for every $\epsilon > 0$ and $R > \rho > 0$, there exists a constant $c = c(\epsilon, n, \rho, R)$ such that*

$$\|Du\|^2_{L^\infty(B_\rho; \mathbb{R}^{N\times n})} \leq c \left\{ \int_{B_R} [1 + f(Du)] \, dx \right\}^{1+\epsilon} . \tag{20}$$

Once we have the estimate (20) for the L^∞*-norm* of the gradient, then the behavior as $t \to +\infty$ of $a(t)$ becomes irrelevant to obtain further regularity of solutions. Therefore we can apply the results already known, with assumptions on the behavior of $a(t)$ as $t \to 0^+$. We refer in particular to the papers [7], [17], [51], [52] for the vector valued case $N \geq 1$ and to [9], [11], [30], [33] for the scalar case $N = 1$. We obtain the following consequence (that can be applied to the examples (6), (7) too).

COROLLARY C *Assume that $a \in C^1((0, +\infty))$ satisfies the nonoscillatory conditions (19), and that there exist an exponent $p \geq 2$ and two positive constants m, M such that*

$$m\, t^{p-2} \leq a(t) \leq a(t) + t \cdot a'(t) \leq M\, t^{p-2},\ \forall t \in (0, 1] \tag{21}$$

(or, equivalently, in terms of the function $g \in C^2([0, +\infty))$,

$$m\, t^{p-2} \leq \frac{g'(t)}{t} \leq g''(t) \leq M\, t^{p-2}\ ,\ \forall t \in (0, 1]\ ); \tag{22}$$

then every weak solution u to the elliptic system (13) *is of class $C^{1,\alpha}_{\text{loc}}(\Omega, \mathbb{R}^N)$.*

If $a(0) > 0$ we are in the case (21) with $p = 2$ and the problem is uniformly elliptic as $t = |\xi| \to 0$. Thus, since $u \in C^{1,\alpha}_{\text{loc}}(\Omega, \mathbb{R}^N)$, each component of the gradient Du is a weak solution to a system with Hölder-continuous coefficients. Then the regularity theory for linear elliptic systems with smooth coefficients applies (see for example Section 3 of Chapter 3 of [16]) and we obtain an higher order regularity result of the type of Corollary B in the previous section.

Acknowledgements. This work is part of the project EURHomogenization, contract SC1-CT91-0732 of the program SCIENCE of the Commission of the European Communities.

References

[1] E. ACERBI & N. FUFSCO, Partial regularity under anisotropic (p, q) growth conditions, *J. Differential Equations*, **107** (1994), 46–67.

[2] T. BHATTACHARYA & F. LEONETTI, A new Poincarè inequality and its applications to the regularity of minimizers of integral functionals with nonstandard growth, *Nonlinear Anal. Theory Methods Appl.*, **17** (1991), 833–839.

[3] T. BHATTACHARYA & F. LEONETTI, $W^{2,2}$ regularity for weak solutions of elliptic systems with nonstandard growth, *J. Math. Anal. Applications*, **176** (1993), 224–234.

[4] L. BOCCARDO, T. GALLOUET & P. MARCELLINI, Anisotropic equations in L^1, *Differential and Integral Equations*, to appear.

[5] L. BOCCARDO, P. MARCELLINI & C. SBORDONE, L^∞-regularity for variational problems with sharp non standard growth conditions, *Boll. Un. Mat. Ital.*, A **4** (1990), 219–225.

[6] M. CHIPOT & L.C. EVANS, Linearization at infinity and Lipschitz estimates for certain problems in the calculus of variations, *Proc. Royal Soc. Edinburgh*, **102** A (1986), 291–303.

[7] H.J. CHOE, Interior behaviour of minimizers for certain functionals with nonstandard growth, *Nonlinear Anal. Theory Methods Appl.*, **19** (1992), 933–945.

[8] E. DE GIORGI, Un esempio di estremali discontinue per un problema variazionale di tipo ellittico, *Boll. Un. Mat. Ital.*, **1** (1968), 135–137.

[9] E. DI BENEDETTO, $C^{1+\alpha}$ local regularity of weak solutions of degenerate elliptic equations, *Nonlinear Anal. Theory Methods Appl.*, **7** (1983), 827–850.

[10] D. M. DUC & J. EELLS, Regularity of exponentially harmonic functions, *International Journal of Math.*, **2** (1991), 395–408.

[11] L.C. EVANS, A new proof of local $C^{1,\alpha}$ regularity for solutions of certain degenerate elliptic p.d.e., *J. Differential Equations*, **45** (1982), 356–373.

[12] N. FUSCO & C. SBORDONE, Higher integrability of the gradient of minimizers of functionals with nonstandard growth conditions, *Comm. Pure Appl. Math.*, **43** (1990), 673–683.

[13] N. FUSCO & C. SBORDONE, Local boundedness of minimizers in a limit case, *Manuscripta Math.*, **69** (1990), 19–25.

[14] N. FUSCO & C. SBORDONE, Some remarks on the regularity of minima of anitropic integrals, *Comm. Partial Differential Equations*, **18** (1993), 153–167.

[15] M. GIAQUINTA, Growth conditions and regularity, a counterexample, *Manuscripta Math.*, **59** (1987), 245–248.

[16] M. GIAQUINTA, Multiple integrals in the calculus of variations and nonlinear elliptic systems, *Annals of Math. Studies* **105**, Princeton Univ. Press, 1983.

[17] M. GIAQUINTA & G. MODICA, Remarks on the regularity of the minimizers of certain degenerate functionals, *Manuscripta Math.*, **57** (1986), 55–99.

[18] E. GIUSTI, Regolarità parziale delle soluzioni di sistemi ellittici quasi lineari di ordine arbitrario, *Ann. Scuola Norm. Sup. Pisa*, **23** (1969), 115–141.

[19] E. GIUSTI, Metodi diretti nel calcolo delle variazioni, *Unione Matematica Italiana*, Bologna, 1994.

[20] E. GIUSTI & M. MIRANDA, Un esempio di soluzioni discontinue per un problema di minimo relativo ad un integrale regolare del calcolo delle variazioni, *Boll. Un. Mat. Ital.*, **2** (1968), 1–8.

[21] E. GIUSTI & M. MIRANDA, Sulla regolarità delle soluzioni deboli di una classe di sistemi ellittici quasi-lineari, *Arch. Rat. Mech. Anal.*, **31** (1968), 173–184.

[22] HONG MIN-CHUNG, Some remarks on the minimizers of variational integrals with non standard growth conditions, *Boll. Un. Mat. Ital.*, A **6** (1992), 91–101.

[23] A.V. IVANOV, Quasilinear degenerate and nonuniformly elliptic and parabolic equations of second order, in "*Proc. Steklov Inst. Math.* ", Russian tom **160** (1982), American Math. Soc., Providence, 1984.

[24] A.G. KOROLEV, On the boundedness of generalized solutions of elliptic differential equations with nonpower nonlinearities, *Mat. Sb.*, **180** (1989), 78–100; *Math. USSR-Sb.*, **66** (1990), 83–106.

[25] M.A. KRASNOSEL'SKII & Y.B. RUTICKI, Convex functions and Orlicz spaces, *Noordhoff Ltd.*, New York, 1961.

[26] O. LADYZHENSKAYA & N. URAL'TSEVA, Linear and quasilinear elliptic equations, in "*Math. in Science and Engineering*," Vol. **46**, Academic Press, San Diego, 1968.

[27] O. LADYZHENSKAYA & N. URAL'TSEVA, Local estimates for gradients of solutions of non-uniformly elliptic and parabolic equations, *Comm. Pure Appl. Math.*, **23** (1970), 677–703.

[28] F. LEONETTI, Two-dimensional regularity of minima of variational functionals without standard growth conditions, *Ricerche di Matematica*, **38** (1989), 41–50.

[29] F. LEONETTI, Weak differentiability for solutions to nonlinear elliptic systems with p,q-growth conditions, *Ann. Mat. Pura Appl.*, **162** (1992), 349–366.

[30] J. LEWIS, Regularity of derivatives of solutions to certain degenerate elliptic equations, *Indiana Univ. Math. J.*, **32** (1983), 849–858.

[31] G.M. LIEBERMAN, The natural generalization of the natural conditions of Ladyzhenskaya and Ural'tseva for elliptic equations, *Commun. Partial Differential Equations*, **16** (1991), 311–361.
[32] G.M. LIEBERMAN, On the regularity of the minimizer of a functional with exponential growth, *Comment. Math. Univ. Carolin.*, **33** (1992), 45–49.
[33] J.J. MANFREDI, Regularity for minima of functionals with p-growth, *J. Differential Equations*, **76** (1988), 203–212.
[34] P. MARCELLINI, Un exemple de solution discontinue d'un problème variationnel dans le cas scalaire, Istituto Matematico "U. Dini," No. 11, Università di Firenze, 1987.
[35] P. MARCELLINI, Regularity of minimizers of integrals of the calculus of variations with non standard growth conditions, *Arch. Rational Mech. Anal.*, **105** (1989), 267–284.
[36] P. MARCELLINI, Regularity and existence of solutions of elliptic equations with p,q-growth conditions, *J. Differential Equations*, **90** (1991), 1–30.
[37] P. MARCELLINI, Regularity for elliptic equations with general growth conditions, *J. Differential Equations*, **105** (1993), 296–333.
[38] P. MARCELLINI, Everywhere regularity for a class of elliptic systems without growth conditions, *Ann. Scuol Norm. Sup. Pisa*, to appear.
[39] P. MARCELLINI, Regularity for some scalar variational problems under general growth conditions, *J. Optim. Theory Appl.*, to appear.
[40] E. MASCOLO & G. PAPI, Local boundedness of minimizers of integrals of the calculus of variations, *Ann. Mat. Pura Appl.*, **167** (1994), 323–339.
[41] E. MASCOLO & G. PAPI, Harnack inequality for minimizers of integral functional with general growth conditions, *Nonlinear Differential Equations Appl.*, to appear.
[42] C.B. MORREY, Partial regularity results for non-linear elliptic systems, *J. Math. and Mech.*, **17** (1968), 649–670.
[43] G. MOSCARIELLO & L. NANIA, Hölder continuity of minimizers of functionals with non standard growth conditions, *Ricerche Mat.*, **40** (1991), 259–273.
[44] J. NEČAS, Example of an irregular solution to a nonlinear elliptic system with analytic coefficients and conditions for regularity, *Theory of Nonlinear Operators, Abhandlungen der Akad. der Wissen. der DDR*, 1977. Proc. of a Summer School held in Berlin in 1975.
[45] J.P. RAYMOND, Lipschitz regularity of solutions of some asymptotically convex problems, *Proc. Royal Soc. Edinburgh*, **117**-A (1991), 59–73.
[46] J. SERRIN, Gradient estimates for solutions of nonlinear elliptic and parabolic equations, in "*Contributions to Nonlinear Functional Analysis*," edited by E.H. Zarantonello, Academic Press, San Diego, 1971, 565–601.
[47] L. SIMON, Interior gradients bounds for non-uniformly elliptic equations, *Indiana Univ. Math. J.*, **25** (1976), 821–855.
[48] B. STROFFOLINI, Global boundedness of solutions of anisotropic variational problems, *Boll. Un. Mat. Ital.*, A **5** (1991), 345–352.
[49] G. TALENTI, Boundedness of minimizers, *Hokkaido Math. J.*, **19** (1990), 259–279.
[50] TANG QI, Regularity of minimizers of nonisotropic integrals in the calculus of variations, *Ann. Mat. Pura Appl.*, **164** (1993), 77–87.
[51] P. TOLKSDORFF, Everywhere regularity for some quasilinear systems with a lack of ellipticity, *Ann. Mat. Pura Appl.*, **134** (1983), 241–266.
[52] K. UHLENBECK, Regularity for a class of non-linear elliptic systems, *Acta Math.*, **138** (1977), 219–240.

Progress in Nonlinear Differential Equations
and Their Applications, Vol. 25

Geodesic Lines in Metric Spaces

MARIO MIRANDA

Dipartimento di Matematica, Università di Trento,
38050 POVO, Trento, Italy

ABSTRACT. The proof of the existence of geodesic lines, given by David Hilbert in 1899 (see D. Hilbert, Ueber das Dirichlet'sche Princip, Jber. Deutsch. Math. Verein., 8 (1900), 184–188), was the first success of the direct method, applied to variational problems.

Hilbert's method works in any metric space, and gives the existence result if the following compactness property is assumed: all closed and bounded subsets of the metric space are compact.

In the first section we expound the method in a general metric space; in the second section we prove that the length of any Lipschitz line can be calculated by a Lebesgue integral.

1 Let M be a metric space, and $(a, b) \in M \times M$. Assume the existence of a continuous line connecting the two points $\underline{a}$ and $\underline{b}$, i.e.

$$\exists\phi : [0,1] \to M, \quad \text{continuous},$$

with

$$\phi(0) = a, \quad \phi(1) = b;$$

and assume

$$|\phi| = \sup\Big\{\sum_{h=0}^{n-1} \text{dist}(\phi(t_h), \phi(t_{h+1})) | n \in N, 0 = t_0 < t_1 < \ldots < t_n = 1\Big\} < +\infty.$$

We can easily modify ϕ to obtain

$$\phi^* : [0, |\phi|] \to M, \quad 1 - \text{Lipschitz},$$

with

$$\phi^*(0) = a, \quad \phi^*(|\phi|) = b;$$

and also

$$\phi^{**} : [0,1] \to M, \quad |\phi| - \text{Lipschitz},$$

with

$$\phi^{**}(0) = a, \phi^{**}(1) = b.$$

ϕ^{**} satisfies also

$$|\phi^{**}| \leq |\phi|.$$

Therefore, the geodesic line problem can be stated as follows: prove the existence of

$$\min\Big\{|\gamma|\big|\gamma\in \mathrm{Lip}([0,1],M),\gamma(0)=a,\gamma(1)=b\Big\}.$$

Hilbert method gives the solution for this problem, if all closed and bounded subset of M are compact.

Proof: Let $\{\gamma_h\}$ be a sequence of lipschitz lines connecting $\underline{a}$ with $\underline{b}$, and such that

$$\lim_h|\gamma_h|=\inf\Big\{|\gamma|\big|\gamma\in \mathrm{Lip}([0,1],M),\gamma(0)=a,\gamma(1)=b\Big\}.$$

Thanks to our previous remark, we can assume that

$$\gamma_h\in \mathrm{Lip}_K([0,1],M),\quad \forall h,$$

for some $K<+\infty$. Thus, there exist a subsequence $\{\gamma_{h(j)}\}$ and

$$\gamma_0\in \mathrm{Lip}_K([0,1],M),$$

such that

$$\lim_j \gamma_{h(j)}(t)=\gamma_0(t),\quad \forall t\in[0,1].$$

From

$$|\gamma_0|\le \liminf_j|\gamma_{h(j)}|$$

we get

$$|\gamma_0|=\min\Big\{|\gamma|\big|\gamma\in \mathrm{Lip}([0,1],M),\gamma(0)=a,\gamma(1)=b\Big\}$$

2 Let

$$\gamma:[0,1]\to M$$

and

$$L=\sup\Big\{\frac{\mathrm{dist}(\gamma(t),\gamma(s))}{|t-s|}\Big|(t,s)\in M\times M, t\neq s\Big\}<+\infty.$$

Define

$$\lambda:[0,1]\to[0,L]$$

by

$$\lambda(s)=\limsup_{t\to s}\frac{\mathrm{dist}(\gamma(t),\gamma(s))}{|t-s|}.$$

We shall prove that λ is Lebesgue-measurable and

$$|\gamma|=\int_0^1\lambda(s)ds.$$

Proof: Define, for $\epsilon>0$ and $s\in[0,1]$,

$$\lambda_\epsilon(s)=\sup\Big\{\frac{\mathrm{dist}(\gamma(t),\gamma(s))}{|t-s|}\Big|t\in[0,1],0<|t-s|<\epsilon\Big\}$$

λ_ϵ is a non-decreasing function of ϵ, and

$$\lim_{\epsilon \to 0} \lambda_\epsilon(s) = \lambda(s), \quad \forall s \in [0,1].$$

It is not difficult to check that each λ_ϵ is a lower semicontinuous function, therefore λ_ϵ is Lebesgue-measurable, and satisfies

$$|\gamma| \leq \int_0^1 \lambda_\epsilon(s)ds.$$

Namely, for a fixed $\epsilon > 0$, if

$$0 = t_0 < t_1 < \ldots < t_n = 1,$$

satisfy

$$t_{h+1} - t_h < \epsilon, \quad \forall h = 0, 1, \ldots, n-1;$$

we get, $\forall \tau \in [t_h, t_{h+1}]$,

$$\operatorname{dist}(\gamma(t_h), \gamma(\tau)) \leq \lambda_\epsilon(\tau)(\tau - t_h),$$
$$\operatorname{dist}(\gamma(t_{h+1}), \gamma(\tau)) \leq \lambda_\epsilon(\tau)(t_{h+1} - \tau),$$

hence

$$\operatorname{dist}(\gamma(t_h), \gamma(t_{h+1})) \leq \lambda_\epsilon(t_{h+1} - t_h),$$

If we choose $\tau_h \in [t_h, t_{h+1}]$, such that

$$\lambda_\epsilon(\tau_h) = \min\{\lambda_\epsilon(\tau) | \tau \in [t_h, t_{h+1}]\},$$

we obtain

$$\sum_{h=0}^{n-1} \operatorname{dist}(\gamma(t_h), \gamma(t_{h+1})) \leq \sum_{h=0}^{n-1} \lambda_\epsilon(\tau_h)(t_{h+1} - t_h) \leq \int_0^1 \lambda_\epsilon(s)ds,$$

and also

$$|\gamma| \leq \int_0^1 \lambda_\epsilon(s)ds,$$
$$|\gamma| \leq \int_0^1 \lambda(s)ds = \lim_{\epsilon \to 0} \int_0^1 \lambda_\epsilon(s)ds.$$

To prove the inequality in the other direction, let us define

$$\ell : [0,1] \to [0, |\gamma|],$$

by

$$\ell(s) = \quad \text{length of} \quad \gamma|_{[0,s]}.$$

ℓ is a Lipschitz function, therefore

$$\ell(1) - \ell(0) = |\gamma| = \int_0^1 \frac{d\ell(s)}{ds} ds.$$

On the other hand

$$\frac{d\ell(s)}{ds} \geq \lambda(s), \text{almost everywhere},$$

thus

$$|\gamma| = \int_0^1 \frac{d\ell(s)}{ds} ds \geq \int_0^1 \lambda(s) ds$$

and

$$|\gamma| = \int_0^1 \lambda(s) ds.$$

Progress in Nonlinear Differential Equations
and Their Applications, Vol. 25

Flow by Mean Curvature of Surfaces of Any Codimension

LUIGI AMBROSIO* AND HALIL METE SONER*

* Istituto di Matematiche Applicate Via Bonanno 25/B, 56126 Pisa, Italy
* Department of Mathematics, Carnegie Mellon University, Pittsburgh, PA 15213, USA

Introduction

In this paper we summarize the main results of [1], where we develop a level set approach for the description of mean curvature flow of surfaces Γ_t of codimension k in $\mathbf{R}^n$, generalizing previous works by Evans & Spruck and Chen, Giga & Goto (see [5], [9]) devoted to the evolution of hypersurfaces.

The main idea, suggested by De Giorgi in [8], is to surround the surface Γ_t by a family of hypersurfaces evolving with normal velocity equal to the sum of the smallest $(n-k)$ principal curvatures. If the hypersurfaces are the level sets of a time depending function u, it turns out that u must satisfy a second order nonlinear, degenerate, parabolic PDE. The theory of viscosity solutions developed in [5], [6], [7], [10] can be applied, yielding existence of a weak solution to the co-dimension k mean curvature flow.

We show the consistency of our weak solutions with classical solutions. The proof is based on the analysis of the properties of the squared distance function η from a smooth manifold Γ and on the relation between the second fundamental form of Γ and the third order derivatives of η on Γ and near to Γ.

Moreover, we compare our level set solution with other solutions already proposed in the literature: the measure theoretic subsolutions of Brakke (see [4]) and Ilmanen (see [12]), the distance solutions of Soner (see [14]) and the minimal barriers of De Giorgi (see [8]).

1. Smooth flow by mean curvature

Let $(\Gamma_t)_{t\in[0,T]}$ be $(n-k)$-dimensional smooth compact manifolds in $\mathbf{R}^n$ without boundary.

DEFINITION. $(\Gamma_t)_{t\in[0,T]}$ *is a smooth flow if there exists a smooth, one to one deformation map* $\phi(\cdot,t):\Gamma_0\to\Gamma_t$ *such that the Jacobian* $J\phi(x,t)$ *is not singular,* $\phi(x,0)=x$ *and* $\phi_t(x,t)$ *is perpendicular to* Γ_t *at* $\phi(x,t)$ *or any* $x\in\Gamma_0$, $t\in[0,T]$. *A smooth flow is a mean curvature flow if*

$$\phi_t(x,t)=\mathbf{H}\big(\phi(x,t),t\big)\qquad\qquad \forall x\in\Gamma_0,\ t\in[0,T]$$

where $\mathbf{H}(y,t)$ *is the mean curvature vector of* Γ_t *at* y.

In the above definition the time derivative $\phi_t(x,t)$ represents the velocity at time t of the point $\phi(x,t) \in \Gamma_t$. The mean curvature vector of a manifold Γ is locally defined by the formula

$$\mathbf{H}(y) := -\sum_{j=1}^{k} \operatorname{div}^{\Gamma}(\nu^j)\nu^j(y) \tag{1.1}$$

where $\nu^1, \ldots, \nu^k$ is a smooth orthonormal vector field generating the normal space to Γ near y and

$$\operatorname{div}^{\Gamma} g := \sum_{i=1}^{n} \delta_i g_i$$

is the tangential divergence. In the co-dimension 1 case, (1.1) reduces to

$$\mathbf{H}(y) := -\operatorname{div}^{\Gamma}(\nu)\nu(y) = -\operatorname{div}(\nu)\nu(y) \tag{1.2}$$

provided ν is a unit vector field defined in a full neighbourhood U of y, perpendicular to Γ at any point in $U \cap \Gamma$. We see from (1.1) that the mean curvature vector $\mathbf{H}$ does not depend on the orientation and is normal to Γ.

Geometrically, $\mathbf{H}$ points in the direction where the $(n-k)$ dimensional area $\mathcal{H}^{n-k}$ of Γ decreases most. This can be seen choosing a vector field $g \in C_0^1(\mathbf{R}^n, \mathbf{R}^n)$, defining

$$\Phi_\tau(x) := x + \tau g(x), \qquad \Gamma_\tau := \Phi_\tau(\Gamma)$$

for $|\tau| << 1$ and looking at the derivative of $\tau \mapsto \mathcal{H}^{n-k}(\Gamma_\tau)$ at $\tau = 0$. Using the divergence theorem on Γ one finds

$$\frac{d}{d\tau}\left(\mathcal{H}^{n-k}(\Gamma_\tau)\right)\bigg|_{\tau=0} = \int_\Gamma \operatorname{div}^{\Gamma} g \, d\mathcal{H}^{n-k} = -\int_\Gamma \langle g, \mathbf{H} \rangle \, d\mathcal{H}^{n-k}.$$

In the case of smooth flow by mean curvature, $\Gamma_{t+\tau} \sim (Id + \tau\mathbf{H}_t)\Gamma_t$ as $\tau \sim 0$, hence

$$\frac{d}{dt}\left(\mathcal{H}^{n-k}(\Gamma_t)\right) = -\int_{\Gamma_t} |\mathbf{H}_t|^2 \, d\mathcal{H}^{n-k}. \tag{1.3}$$

It it well known that the flow by mean curvature may develop singularities in finite time even if the initial surface is smooth, making meaningless (with the exception of simple plane curves, boundaries of convex sets and graphs) the parametric approach. This is the main motivation for the research of a weak definition of flow by mean curvature, consistent with the classical one and defined for all times, even after the appearance of singularities.

2. The level set approach

The basic idea of the level set approach (see [13], [9], [5]) is to find a parabolic PDE such that *all* level sets of any solution u flow by mean curvature. The crucial property is that

$$t \mapsto \operatorname{dist}\big(\Gamma_t, \Gamma_t'\big)$$

is non decreasing for hypersurfaces Γ_t, Γ_t' flowing by mean curvature, so that initially disjoint hypersurfaces remain disjoint, and we can view them as level sets of a time depending function u. We introduce the following notations

$$\nu := \frac{\nabla u}{|\nabla u|} \qquad\qquad P_w := I - \frac{w \otimes w}{|w|^2} \quad w \neq 0.$$

Hence, ν is a unit normal vector to the level sets of u and P_w is the orthogonal projection on the hyperplane normal to w.

Using (1.2) and the fact that the velocity (in the direction ν) of a point $x \in \mathbf{R}^n$ is given by $-u_t(x,t)/|\nabla u|(x,t)$, we find that the PDE is

$$\begin{aligned} u_t &= |\nabla u| \operatorname{div}\left(\frac{\nabla u}{|\nabla u|}\right) = \big(\delta_{ij} - \nu_i \nu_j\big) \frac{\partial^2 u}{\partial x_i \partial x_j} \\ &= F\big(\nabla u, \nabla^2 u\big) \end{aligned} \tag{2.1}$$

with $F(w,X) := \operatorname{trace}\big(P_w X\big)$ (here and in the following we use the summation convention on repeated indices). Generalized solutions of the mean curvature flow are defined in the following steps:

1) Find u_0 such that $\Gamma_0 = \big\{x : u_0(x) = 0\big\}$.
2) Solve the PDE with the initial condition $u(x,0) = u_0(x)$.
3) Define $\Gamma_t := \big\{x : u(x,t) = 0\big\}$.
4) Show that Γ_t depends only on Γ_0, i.e., it does not depend on the choice of u_0.

However, there are some difficulties: the equation $u_t = F\big(\nabla u, \nabla^2 u\big)$ is non-linear, degenerate and the 0-homogeneous map $w \mapsto F(w,X)$ has no continuous extension at 0. All these difficulties can be removed using the theory of viscosity solutions (see [6], [7]). Let us recall the definition of viscosity solution of a second order parabolic *PDE*:

Definition. *Let $u : \mathbf{R}^n \times (0,T) \to \mathbf{R}$ be a continuous function, let A be a dense subset of $\mathbf{R}^n \times \mathbf{S}^{n\times n}$ and $F : A \to \mathbf{R}$.*

u subsolution of $u_t = F\big(\nabla u, \nabla^2 u\big)$ $\iff$ *$u - \phi$ relative maximum at (x_0, t_0), ϕ of class C^2, implies $\phi_t \le F^*\big(\nabla\phi, \nabla^2\phi\big)$ at (x_0,t_0)*

u supersolution of $\iff$ *$u - \phi$ relative minimum at (x_0, t_0), ϕ of class C^2, implies*

$u_t = F(\nabla u, \nabla^2 u)$ $\qquad\qquad$ $\phi_t \ge F_*(\nabla\phi, \nabla^2\phi)$ at (x_0, t_0)

u viscosity solution $\iff$ *u subsolution and supersolution.*

In the above definition $\mathbf{S}^{n\times n}$ denotes the set of symmetric $n \times n$ matrices and F^*, F_* denote the upper and lower semicontinuous extensions of F

$$F^*(w, X) := \sup\left\{\limsup_{h\to+\infty} F(w_h, X_h) : (w_h, X_h) \in A, \quad (w_h, X_h) \to (w, X)\right\}$$

$$F_*(w, X) := \inf\left\{\liminf_{h\to+\infty} F(w_h, X_h) : (w_h, X_h) \in A, \quad (w_h, X_h) \to (w, X)\right\}$$

defined for all pairs $w \in \mathbf{R}^n$ and $X \in \mathbf{S}^{n\times n}$.

The basic assumption of the theory of viscosity solutions is the *degenerate ellipticity* of F_* and F^*:

$$X \ge Y \quad \Longrightarrow \quad F_*(w, X) \ge F_*(w, Y), \quad F^*(w, X) \ge F^*(w, Y).$$

The degenerate ellipticity implies that classical solutions of the equation are viscosity solutions. Indeed, if $u \in C^2(\mathbf{R}^n \times (0, T))$ is a classical solution of the equation and $u - \phi$ has a relative maximum at (x_0, t_0), then $\nabla u = \nabla\phi$, $u_t = \phi_t$, $\nabla^2 u \le \nabla^2\phi$ at (x_0, t_0), hence

$$\begin{aligned} F^*\left(\nabla\phi(x_0, t_0), \nabla^2\phi(x_0, t_0)\right) &= F^*\left(\nabla u(x_0, t_0), \nabla^2\phi(x_0, t_0)\right) \\ &\ge F^*\left(\nabla u(x_0, t_0), \nabla^2 u(x_0, t_0)\right) \\ &= u_t(x_0, t_0) = \phi_t(x_0, t_0) \end{aligned}$$

and u is a viscosity subsolution. A similar argument works for supersolutios. The main advantages of viscosity solutions are:

1) General existence results, by Perron's method (see [6], [7], [5]).
2) Comparison theorems (see [10]):

$$\left.\begin{array}{l} u \text{ subsolution in } \mathbf{R}^n \times (0, T) \\ v \text{ supersolution in } \mathbf{R}^n \times (0, T) \\ \\ u,\ v \text{ uniformly continuous} \\ u(x, 0) \le v(x, 0) \text{ for any } x \in \mathbf{R}^n \end{array}\right. \quad \Longrightarrow \quad u \le v \text{ in } \mathbf{R}^n \times (0, T).$$

3) Strong stability properties with respect to uniform convergence (even Γ-convergence) of F and/or u (see [5]).

Evans & Spruck in [9], and Chen, Giga & Goto in [5] independently proved existence and uniqueness of viscosity solutions of

$$\begin{cases} u_t = F(\nabla u, \nabla^2 u) \\ u(x, 0) = u_0(x). \end{cases}$$

Moreover, they proved that

$$\Gamma_t := \{x : u(x,t) = 0\}$$

depends only on Γ_0 and that the smoothness of Γ_0 implies the coincidence of the level sets Γ_t with the *classical* evolution of Γ_0, as long as the latter is defined. Moreover, short time existence results for the classical evolution starting from Γ_0 have been obtained by De Mottoni & Schatzman, Huisken, Evans & Spruck.

3. Higher codimension level set flow

The extension of the level set approach to the description of higher codimension flows is not straightforward. The main difficulty is due to the fact that a codimension k manifold can be represented by the intersection of k level sets of scalar functions. Hence, the level set approach leads to a *system* of PDE, without maximum principle. Geometrically, the absence of a maximum principle is related to the fact that initially disjoint manifolds could intersect during their flow by mean curvature. This happens, for instance, if Γ_0 consists of two chained rings. An additional difficulty is related to the computation of $\mathbf{H}$, requiring by (1.1) the local knowledge of an orthonormal basis of the manifold.

The basic idea, due to De Giorgi, is to look at the evolution of all positive level sets of a *scalar* function $u \geq 0$ under a new geometric law. Specifically, the level sets are required to flow with velocity equal to the sum of the smallest $(n-k)$ principal curvatures. Quite surprisingly, it turns out that the 0 level set of u is forced to flow by mean curvature!
The necessity of removing the highest $(k-1)$ eigenvalues to get a nonconstant motion can be understood by looking at the evolution of a simple smooth curve Γ in $\mathbf{R}^3$. In this case, if we look at a small tubular surface S around Γ, we find that one of the principal curvatures is very high, its order of magnitude being the inverse of $\mathrm{dist}(S,\Gamma)$, and the other one is related to the geometry of Γ. In particular, if we let S flow by mean curvature, we find that S becomes empty in a very short time.

Let u be a nonnegative function, assume that $u \in C^2(\{u > 0\})$ and $|\nabla u| > 0$ for $u > 0$. For $\tau > 0$, let

$$E^\tau := \{x : u(x) = \tau\}$$

oriented da ν. The principal curvatures of E^τ (with respect to the orientation induced by ν) are given by the eigenvalues of the symmetric bilinear form

$$B(\xi,\eta) := \langle \xi, d_\eta \nu^{E^\tau} \rangle \qquad \xi,\ \eta \in \nu^\perp.$$

With this sign convention, if u is a convex function, the sets $\{x : u(x) < \tau\}$ have nonnegative principal curvatures. A simple computation shows that

$$B = \frac{P_\nu \nabla^2 u P_\nu}{|\nabla u|} \qquad \text{on } \nu^\perp. \tag{3.1}$$

Hence, we define

$$\begin{aligned} G(w,X) : &= \text{sum of the } (n-k) \text{ smallest eigenvalues} \\ &\quad \text{of } Y := P_w X P_w \text{ on the space } w^\perp \\ &= \lambda_1(Y) + \ldots + \lambda_{n-k}(Y) \end{aligned} \tag{3.2}$$

removing the highest $(k-1)$ eigenvalues of Y.
By (3.1), (3.2), the positive level sets of a function $u(x,t)$ flow (in the direction $-\nu$) with velocity equal to the sum of their smallest $(n-k)$ principal curvatures if and only if

$$u_t = G\big(\nabla u, \nabla^2 u\big), \qquad u(x,0) = u_0(x) \geq 0. \tag{3.3}$$

To our knowledge, there are no results in the literature concerning short time existence of $C^{1,1}$ solutions of (3.3), assuming that $u_0 \in C^2\big(\{u_0 > 0\}\big)$ and $|\nabla u_0| > 0$ for $u_0 > 0$. Even if u_0 is smooth, we cannot expect smooth solutions, even for a short time, because $X \mapsto G(w,X)$ is not a C^1 function (only a Lipschitz function) in $\mathbf{S}^{n\times n}$.
However, the theory of viscosity solutions can be used to get weak solutions of (3.3). In order to apply this theory, we need only to check the degenerate ellipticity of $X \mapsto G(w,X)$. The problem is the following: given a $(n-1)$-dimensional vector space E, a symmetric bilinear form Y on E, and given the ordered list of the eigenvalues of Y

$$\lambda_1(Y) \leq \lambda_2(Y) \leq \ldots \leq \lambda_{n-1}(Y)$$

can we say that

$$Y \leq Y' \qquad \Longrightarrow \qquad \lambda_i(Y) \leq \lambda_i(Y') \quad i = 1, \ldots, (n-1) \quad ?$$

The answer is positive because the numbers $\lambda_i(Y)$ solve a max-min problem depending monotonically on Y:

$$\lambda_i(Y) = \max\left\{ \min_{p \in F\setminus\{0\}} \frac{\langle Yp, p\rangle}{|p|^2} : F \subset E,\ \operatorname{codim} F \leq (i-1) \right\}. \tag{3.4}$$

The proof of (3.4) is elementary. Indeed, the inequality $\leq$ follows by choosing as F the vector space spanned by the eigenvectors corresponding to $\lambda_i(Y)$, $\lambda_{i+1}(Y)$, $\ldots$, $\lambda_n(Y)$. The opposite inequality follows by the fact that each subspace F with codimension not greater than $(i-1)$ has at least a nonzero vector in common with the vector space F_0 generated by the eigenvectors corresponding to $\lambda_1(Y) \ldots, \lambda_i(Y)$. Since

$$Y \leq \lambda_i(Y) I \quad \text{on } F_0 \qquad \text{and} \qquad Y \geq \min_{p \in F\setminus\{0\}} \frac{\langle Yp, p\rangle}{|p|^2} I \quad \text{on } F$$

we obtain

$$\lambda_i(Y) \geq \min_{p \in F\setminus\{0\}} \frac{\langle Yp, p\rangle}{|p|^2}.$$

Hence, $G(w, X) = \lambda_1(Y) + \ldots + \lambda_{n-k}(Y)$ is degenerate elliptic. Geometrically, the ellipticity of G is related to the validity of an inclusion principle for open sets whose boundaries flow by the sum of the smallest $(n-k)$ principal curvatures, i.e.,

$$\Omega(t_0) \subset \Omega'(t_0) \qquad \Longrightarrow \qquad \Omega(t) \subset \Omega'(t) \quad \forall t \geq t_0.$$

Therefore, the level set approach can be used to describe this motion. The surprising property is that the flow according to this law of a family of hypersurfaces Γ_t^τ filling $\mathbf{R}^n \setminus \Gamma_t$ forces Γ_t to flow by mean curvature.

Applying a general existence theorem of Chen, Giga & Goto for "geometric" parabolic equations (see [5]) we proved that

$$\begin{cases} u_t = G(\nabla u, \nabla^2 u) \\ u(x,0) = u_0(x) \end{cases} \tag{3.5}$$

has a unique viscosity solution for any uniformly continuous function u_0.

DEFINITION. *Given a closed set Γ_0, we define generalized co-dimension k flow of Γ_0 the sets*

$$\Gamma_t := \{x : u(x,t) = 0\}$$

where $u(x,t)$ is the solution of the problem (3.5) with a nonnegative, uniformly continuous function u_0 such that

$$\Gamma_0 = \{x : u_0(x) = 0\}.$$

The definition is well posed: using comparison theorems we proved that Γ_t does not depend on the choice of u_0. In the case $k = 1$ it is easy to check that G is equal to the function F considered by Evans & Spruck, Chen, Giga & Goto. Since the map $t \mapsto u(t, \cdot)$ has the semigroup property, we have the

Semigroup property: Given s, $t \geq 0$, the set Γ_{t+s} coincides with the evolution at time t starting from Γ_s.

4. Main properties of co-dimension k flow

The co-dimension k level set flow Γ_t defined in §3 can be compared with the classical solutions and other weak solutions of the flow by mean curvature proposed by Brakke, Ilmanen, De Giorgi. It turns out that, if Γ_0 is smooth, Γ_t coincides with the classical solution of flow by mean curvature as long as the latter is defined. We will explain in the next section the ideas involved in the proof of this consistency result.

Let us very briefly recall Brakke's definition of motion by mean curvature. A family of Radon measures $(\mu_t)_{t \geq 0}$ in $\mathbf{R}^n$ is said to be a *m-dimensional Brakke motion* if

$$\frac{d^+}{dt}\left(\int_{\mathbf{R}^n} \phi \, d\mu_t\right) \leq \mathcal{B}(\mu_t, \phi) \qquad \forall \phi \in C_0^2((\mathbf{R}^n, [0, +\infty)) \tag{4.1}$$

where d^+/dt denotes the upper derivative and

$$\mathcal{B}(\mu_t, \phi) = \int_{\mathbf{R}^n} \left(-\phi|\mathbf{H}_t|^2 + \langle \nabla\phi, \mathbf{H}_t\rangle\right) d\mu_t \tag{4.2}$$

if μ_t is equal to a m-varifold with integer density and absolutely continuous mean curvature $H_t\mu_t$ in $\{\phi > 0\}$ (see [4], [12]) and $\mathcal{B}(\mu_t, \phi) = -\infty$ otherwise. For a smooth mean curvature flow, (4.2) holds and the equality in (4.1) can be proved by a localization of (1.3).

As Ilmanen proved in [12], we can think to the level set flow as a minimal set theoretic supersolution and to the Brakke's flow as a measure theorethic subsolution, because only the inequality $\leq$ is required in (4.1). Hence, the best result we can hope for is an inclusion property of $(n-k)$-dimensional Brakke's motions in co-dimension k level set motions. Indeed for any Brakke motion (see [1], Theorem 5.4) the following implication holds:

$$\operatorname{supp}\mu_0 \subset \Gamma_0 \qquad \Longrightarrow \qquad \operatorname{supp}\mu_t \subset \Gamma_t \quad \forall t \geq 0.$$

Similar comparison properties can be stated and proved for distance solutions (see [14] and §4 of [1]) and for the barrier solutions introduced by De Giorgi (see [2], [3], [8] and §6 of [1]).

5. Consistency with smooth flows

The typical idea to prove the consistency result (already exploited in codimension 1) is to describe the flow by mean curvature using a PDE satisfied by the distance function, and use this to compare classical and viscosity solutions. In the case $k = 1$ it is well known that

$$\begin{gathered}\Gamma_t = \partial\Omega_t \text{ is a smooth mean curvature flow}\\ \Updownarrow \\ r_t = \Delta r \quad \text{on} \quad \{r = 0\}\end{gathered}$$

where $r(x,t)$ is the signed distance function from Γ_t:

$$r(x,t) = \begin{cases} -\operatorname{dist}(x, \Gamma_t) & \text{if } x \in \Omega_t;\\ \operatorname{dist}(x, \Gamma_t) & \text{if } x \notin \Omega_t.\end{cases}$$

In higher codimension there is no possibility to define a signed distance function, so that the first idea could be to work with the distance function

$$\delta(x) := \operatorname{dist}(x, \Gamma).$$

However, δ is a Lipschitz function in $\mathbf{R}^n$ but its first order derivatives (defined in $\{0 < \delta < \tau\}$ for small τ) are discontinuous on Γ. Moreover $(k-1)$ eigenvalues of the hessian matrix of δ are unbounded near Γ and even if we remove them there is still lack of continuity. Indeed, we proved that

$$\nexists \lim_{y \to x \in \Gamma} G\left(\nabla\delta(y), \nabla^2\delta(y)\right)$$

because the limit above exists only on lines normal to Γ, depends on the direction $p \in \mathbf{S}^{n-1}$ of the line and is equal to $-\langle \mathbf{H}(x), p\rangle$.

As suggested by De Giorgi, to transfer informations from Γ to a neighbourhood of Γ and to characterize the flow by mean curvature we will use the *squared distance function*

$$\eta(x) := \frac{1}{2}\delta^2(x) = \frac{1}{2}\mathrm{dist}^2(x,\Gamma).$$

The first result we proved (see [1], Theorem 3.5 and Lemma 3.7) is the following:

THEOREM 1. *Let $\{\Gamma_t\}_{t\in[0,T]}$ be a smooth flow of codimension k and let*

$$\eta(x,t) := \mathrm{dist}^2(x,\Gamma_t)/2.$$

Then, Γ_t is a mean curvature flow if and only if

$$(\nabla\eta)_t = \Delta(\nabla\eta) \qquad \text{on } \{\eta = 0\}.$$

The proof of Theorem 1 is based on the fact that $-(\nabla\eta)_t(x,t)$ is the normal velocity of the point $x \in \Gamma_t$ and $-\Delta(\nabla\eta(x,t))$ is the mean curvature vector $\mathbf{H}_t(x)$ of Γ_t at x. Hence, one of the advantages of the function η is that it provides a simple method for the computation of the mean curvature vector.

The second result (see [1], Theorem 3.8) shows that the system in η is equivalent to a differential inequality in δ, where

$$\delta(x,t) := \sqrt{2\eta} = \mathrm{dist}(x,\Gamma_t).$$

THEOREM 2. *Let $\{\Gamma_t\}_{t\in[0,T]}$ be a smooth flow of codimension k, and let $\Omega \subset \mathbf{R}^n \times (0,T)$ be the maximal open set where δ is smooth. Then, Γ_t flows by mean curvature if and only if*

$$\delta_t \geq G(\nabla\delta, \nabla^2\delta) \qquad \text{on } \Omega.$$

The proof of Theorem 2 also shows that the inequality

$$\delta_t \geq G(\nabla\delta, \nabla^2\delta) \tag{5.1}$$

holds in $\mathbf{R}^n \times (0,T)$, in the viscosity sense. Moreover,

$$\delta_t \leq G(\nabla\delta, \nabla^2\delta) + C\delta \qquad \text{on } \{0 < \delta < \sigma\} \tag{5.2}$$

for suitable constants C, $\sigma > 0$. From (5.2) we get that $W := e^{-Ct}(\delta \wedge \sigma/2)$ satisfies

$$W_t \leq G(\nabla W, \nabla^2 W) \tag{5.3}$$

in $\mathbf{R}^n \times (0,T)$, in the viscosity sense. Hence, using (5.1) and (5.3) we can compare δ with the viscosity solution of

$$\begin{cases} u_t = G(\nabla u, \nabla^2 u) \\ u(x,0) = \delta(x) \end{cases}$$

obtaining $W \leq u \leq \delta$ in $\mathbf{R}^n \times (0,T)$. In particular, since W and δ have the same 0 level set,

$$\Gamma_t = \{x : \delta(x,t) = 0\} = \{x : u(x,t) = 0\} \qquad \forall t \in (0,T)$$

and this shows the consistency with smooth flows.

The proof of Theorem 2 is based on the propagation properties of the eigenvalues of $\nabla^2\eta$ on normal lines to Γ. Given a smooth, compact, codimension k surface Γ without boundary, let $\tau > 0$ such that $\eta \in C^\infty(\{\eta \leq \tau^2/2\})$ and let $x \notin \Gamma$ such that $\delta(x) = \tau$. Denoting by $y \in \Gamma$ the point in Γ of least distance from x, set

$$p := \frac{x-y}{\tau}, \qquad x_s := y + sp, \qquad B(s) := \nabla^2\eta(x_s)$$

for $s \in [0,\tau]$. Then, the following theorem holds (see [1], Theorem 3.2):

THEOREM 3. *(1) For $s \in [0,\tau]$ the eigenvectors of $B(s)$ do not depend on s, $B(s)$ has k eigenvalues equal to 1, all others $\lambda_i(s)$ are strictly less than 1 and satisfy*

$$|\lambda_i(s)| \leq Cs \qquad i = 1,\ldots,n-k$$

with C depending only on Γ, τ.
(2) For $s \in (0,\tau]$ the matrix $\nabla^2\delta(x_s)$ has $(k-1)$ eigenvalues equal to $1/s$, one (corresponding a p) equal to 0, all others $\beta_i(s)$ are strictly less than $1/s$ and satisfy

$$|\beta_i(s)| \leq C \qquad i = 1,\ldots,n-k.$$

(3) The maps

$$s \mapsto \beta_i(s), \qquad s \mapsto G(\nabla\delta(x_s), \nabla^2\delta(x_s))$$

are nonincreasing in $(0,\tau]$.

The geometric interpretation of Theorem 3 is the following: consider, for instance, the level set $E_s := \{\delta = s\}$ of the distance function δ from a smooth, simple curve Γ in $\mathbf{R}^n$. Then, if E_s lies in the region where δ^2 is smooth, we know that $(n-2)$ principal curvatures of E_s are *exactly* equal to the $1/s$.

The proof of Theorem 3 is based on the continuity of the map $s \mapsto B(s)$ in $[0,\tau]$ and on the fact that $B(s)$ solves the ODE

$$B'(s) = \frac{B(s) - B^2(s)}{s} \qquad s \in (0,\tau]. \tag{5.4}$$

The ODE is obtained by differentiating twice $|\nabla\eta|^2 = 2\eta$ (this identity directly follows from the equality $|\nabla\delta|^2 = 1$). Indeed, the first differentiation gives $\eta_{ij}\eta_j = \eta_i$ and the second one gives

$$\eta_{ijk}\eta_j + \eta_{ij}\eta_{jk} = \eta_{ik}. \tag{5.5}$$

It is not hard to see that $\eta_{ijk}(x_s)\eta_j(x_s)$ is equal to $B'_{ij}(s)s$, hence (5.4) follows by (5.5). The ODE implies that the matrices $B(s)$ are diagonal in a common basis and their eigenvalues satisfy

$$\lambda'_i(s) = \frac{\lambda_i(s) - \lambda_i^2(s)}{s}. \tag{5.6}$$

On the other hand, $B(0)$ is the orthogonal projection on the normal space to Γ at y, hence $B(0)$ has k eigenvalues equal to 1. Studying the solutions of (5.6) backward in time we obtain that $B(s)$ must have exactly k eigenvalues equal to 1 and the other ones must be infinitesimal (with order s) as $s \to 0^+$. The connection between the eigenvalues of $\nabla^2\delta$ and the eigenvalues of $\nabla^2\eta$ comes from the fact that $\nabla\delta$ is an eigenvalue of $\nabla^2\eta$ and from the identity

$$\delta\delta_{ij} = \eta_{ij} - \delta_i\delta_j$$

that can be obtained by differentiation of $\eta_i = \delta\delta_i$.

Now we can explain the connection, stated in Theorem 2, between the differential inequality

$$\delta_t \geq G\big(\nabla\delta, \nabla^2\delta\big) \qquad \text{on} \quad \big\{0 < \delta \leq \tau\big\}$$

and the system

$$\big(\nabla\eta\big)_t = \Delta\big(\nabla\eta\big) \qquad \text{on} \quad \big\{\eta = 0\big\}.$$

Let us set

$$\alpha_i := (\eta_i)_t - \Delta(\eta_i), \qquad \beta_1 \leq \beta_2 \leq \ldots \leq \beta_n \quad \text{eigenvalues of } \nabla^2\delta.$$

By Theorem 3, $\beta_{n-k+2} = \cdots = \beta_n = \delta^{-1}$. Using again identities obtained by differentiation of $|\nabla\delta|^2 = 1$ we get the following identities

$$\begin{aligned}
\alpha_i\delta_i &= \delta_t - \Delta\delta + \delta\|\nabla^2\delta\|^2 \\
&= \delta_t - \sum_{i=1}^{n}\beta_i + \delta\sum_{i=1}^{n}\beta_i^2 && \text{(by the symmetry of } \nabla^2\delta) \\
&= \delta_t - \sum_{i=1}^{n-k+1}\beta_i + \delta\sum_{i=1}^{n-k+1}\beta_i^2 && \text{(by Theorem 3)} \\
&= \delta_t - G\big(\nabla\delta, \nabla^2\delta\big) + O(\delta) && \text{(by Theorem 3).}
\end{aligned}$$

in the set

$$\big\{(x,t) : 0 < \delta(x,t) \leq \tau\big\}.$$

Assume that $\alpha_i = 0$ for $\eta = 0$. Passing to the limit we get

$$\lim_{x \to y \in \Gamma} \delta_t - G\big(\nabla\delta, \nabla^2\delta\big) = 0.$$

Since $\delta_t(x,t)$ is constant and $G\big(\nabla\delta, \nabla^2\delta\big)$ is nonincreasing on normal lines to Γ (moving away from Γ) we infer

$$\delta_t \geq G\big(\nabla\delta, \nabla^2\delta\big) \qquad \text{on} \quad \big\{0 < \delta \leq \tau\big\}.$$

Conversely, assume that the differential inequality holds. Approaching to Γ on a line parallel to $p \in \mathbf{S}^{n-1}$, since $\delta_i = p_i$ on the line, we get

$$\alpha_i p_i \geq 0 \qquad \text{on} \quad \big\{\eta = 0\big\}.$$

Since $\alpha = \mathbf{H} - V$ is normal to Γ and p is arbitrary, $\alpha = 0$.

References

[1] **L. Ambrosio and H.M. Soner:** *Level set approach to mean curvature flow in arbitrary codimension.* Submitted to J. Diff. Equations, 1994.

[2] **G. Bellettini and M. Paolini:** *Some results on minimal barriers in the sense of De Giorgi applied to motion by mean curvature.* Atti Accad. Naz. Lincei, Cl. Sci. Fis. Mat. Natur. (1994), to appear.

[3] **G. Bellettini and M. Paolini:** *Some results on De Giorgi's notion of minimal barrier applied to motion by mean curvature.* Atti Accad. Naz. XL, Cl. Sci. Fis. Mat. Natur. (1994), to appear.

[4] **K.A. Brakke:** *The motion of a surface by its mean curvature.* Princeton University Press, Princeton, 1978.

[5] **Y.-G. Chen, Y. Giga and S. Goto:** *Uniqueness and existence of viscosity solutions of generalized mean curvature flow equations.* J. Diff. Geom., **33** 749–786, (1991).

[6] **M.G. Crandall, L.C. Evans and P.L. Lions:** *Some properties of viscosity solutions of Hamilton-Jacobi equations.* Trans. AMS, **282** 487–502, (1984).

[7] **M.G. Crandall, H. Ishii and P.L. Lions:** User's guide to viscosity solutions of second order partial differential equations. Bull. AMS, **27** 1–67, (1992).

[8] **E. De Giorgi:** *Barriers, boundaries, motion of manifolds.* Lectures held in Pavia, 1994.

[9] **L.C. Evans and J. Spruck:** Motion of level sets by mean curvature I, II, III, IV J. Diff. Geom., **33**, 635–681 (1991), Trans. AMS, **330**, 321–332 (1992), J. Geom. Analysis, **2**, 121–150 (1992), J. Geom. Analysis (1993), to appear.

[10] **Y. Giga, S. Goto, H. Ishii and M.H. Sato:** *Comparison principle and convexity preserving properties for singular degenerate parabolic equations on unbounded domains.* Indiana Univ. Math. J. **40** 443–470 (1992).

[11] **G. Huisken:** *Flow by mean curvature of convex surfaces into spheres.* J. Diff. Geom., **20** 237–266, (1984).

[12] **T. Ilmanen:** *Elliptic regularization and partial regularity for motion by mean curvature.* Memoirs AMS **108** (1994).

[13] **S. Osher and J. Sethian:** *Fronts propagating with curvature depending speed.* J. Comp. Phys. **79** 12–49, (1988).

[14] **H.M. Soner:** *Motion of a set by the curvature of its boundary.* J. Diff. Equations, **101** 313–372, (1993).

Progress in Nonlinear Differential Equations
and Their Applications, Vol. 25

Functions of Bounded Variation over Non-Smooth Manifolds and Generalized Curvatures

GABRIELE ANZELLOTTI

Dipartimento di Matematica, Università di Trento
38050 POVO, Trento, Italy

ABSTRACT. The aim of this paper is to describe some recent attempts to introduce a reasonable notion of a function of bounded variation on a non-smooth manifold. This investigation is motivated by
a) variational problems on a fixed non-smooth domain;
b) problems where the domain is itself to be determined (hence a priori possibly non regular);
c) problems involving curvatures (for surfaces with curvature measures one expects the normal vector to be of bounded variation).
As non-smooth manifolds we choose to take rectifiable sets and rectifiable currents. We shall use currents and integration by parts as a basic tool.

1. BV functions on a rectifiable current

1.1. BV functions on a euclidean space. If Ω is an open set in $\mathbf{R}^n$ and $u : \Omega \to \mathbf{R}$ is a Lebesgue summable function, then u is said to be a function of bounded variation ($u \in BV(\Omega)$) if the distributional gradient Du of u is a vector measure of bounded total variation in Ω. In other words, $u \in BV(\Omega)$ if and only if there are real measures of bounded variation μ_i in Ω ($i = 1, \ldots, n$) such that

$$\int_\Omega u D_i \varphi dx = - \int_\Omega \varphi d\mu_i \qquad \forall \varphi \in C_o^\infty(\Omega), \forall i = 1, \ldots, n.$$

If $u \in BV(\Omega)$ then the total variation of Du in Ω is

$$|Du|(\Omega) = \sup\Big\{ \int_\Omega \sum_{i=1}^n u D_i \varphi_i \quad | \quad \varphi_i \in C_o^\infty(\Omega)\,,\ \sum_i \varphi_i(x)^2 \le 1 \Big\}.$$

1.2. Integration by parts on a rectifiable current.
Notation:

M	is a rectifiable set of dimension k in $\mathbf{R}^n$, of finite $\mathcal{H}^k$ Hausdorff measure,
$T_x M$	is the tangent plane to M at x, defined $\mathcal{H}^k$-a.e. on M,
$\tau(x)$	is a measurable field of unit simple k-vectors associated to $T_x M$.

We consider the k-dimensional rectifiable current $S = [\![M, \tau, 1]\!]$ defined by

$$\langle S, \omega \rangle = \int_M \langle \tau(x), \omega(x) \rangle d\mathcal{H}^k$$

for all smooth and compactly supported k-forms ω in $\mathbf{R}^n$. If $\varphi \in C_o^\infty(\mathbf{R}^n)$ one has immediately

$$\partial(\varphi S) = \varphi \partial S - S \llcorner d\varphi.$$

If ∂S is also rectifiable, say $[\![\Gamma, \zeta, 1]\!]$, then one has

$$\langle \partial(\varphi S), dx^\alpha \rangle = \int_\Gamma \varphi \zeta^\alpha d\mathcal{H}^{k-1} - \int_M \langle \tau(x) \llcorner d\varphi(x), dx^\alpha \rangle d\mathcal{H}^k \tag{1.2.1}$$

for all $\alpha \in \Lambda(n, k-1)$. Hence, because $\langle \partial(\varphi S), dx^\alpha \rangle = 0$, we obtain the integration by parts formula

$$\int_\Gamma \varphi \zeta^\alpha d\mathcal{H}^{k-1} = \int_M R_\tau^\alpha(\varphi) d\mathcal{H}^k \qquad \forall \alpha, \forall \varphi \tag{1.2.2}$$

where we have set

$$R_\tau^\alpha(\varphi) = \langle \tau(x) \llcorner d\varphi(x), dx^\alpha \rangle.$$

The $(k-1)$-vector $R_\tau(\varphi) = \tau(x) \llcorner d\varphi(x)$ is a *tangential gradient* of φ at x.

1.3. BV_* **functions on a rectifiable current.** DEFINITION 1.3.1 *We say that a function $u : M \to \mathbf{R}$ is in $BV_*(\Omega)$ if u is $\mathcal{H}^k$-summable and $\|\partial(uS)\| < +\infty$.*

The definition above is modelled after the one in Federer's book, and reduces to the classical notion given in 1.1 if $k = n$ and $M = \Omega$.
Both the current $\partial(uS)$ and its mass $\|\partial(uS)\|$ may be characterized by letting u act suitably on the tangential gradient of test functions on M. Precisely, we have

$$\langle \partial(uS)^\alpha, \varphi \rangle = \int_M u \langle \tau \llcorner d\varphi, dx^\alpha \rangle d\mathcal{H}^k = \int_M u R_\tau^\alpha(\varphi) d\mathcal{H}^k \qquad \forall \alpha$$

and

$$\|\partial(uS)\| = \sup\Big\{ \int_M u R_\tau^\alpha(\varphi_\alpha) d\mathcal{H}^k \;\Big|\; \varphi_\alpha \in C_o^\infty(\mathbf{R}^n) \,,\, \sum_\alpha \varphi_\alpha(x)^2 \le 1 \Big\}. \tag{1.3.1}$$

We remark that u is in $BV_*(\Omega)$ if and only if $R_\tau^\alpha : \varphi \to R_\tau^\alpha(\varphi)$ is a bounded measure for all α. In this case the number $\|\partial(uS)\|$ is equal to the total variation of the vector measure $R_\tau : \varphi \to R_\tau(\varphi)$, which essentially coincides to Du in the case that $k = n$ and $M = \Omega$ (compare with [ADS]).

1.4. The graph of a function over a rectifiable current. With the notation of section 1.2 we consider the rectifiable set

$$E_u = \big\{ (x,t) \in M \times \mathbf{R} \mid 0 < t < u(x) \text{ if } u(x) > 0 \,,\; 0 > t > u(x) \text{ if } u(x) < 0 \big\}.$$

For $\mathcal{H}^{k+1}$-almost all $(x,t) \in E_u$ we consider the $(k+1)$-vector

$$\alpha(x,t) = \begin{cases} -\varepsilon \wedge \tau(x) & \text{if } u(x) > 0; \\ \varepsilon \wedge \tau(x) & \text{if } u(x) < 0. \end{cases}$$

DEFINITION 1.4.1. [ADS] *We say that the* $(k+1)$*-rectifiable current* $\Sigma_u = [\![E_u, \alpha, 1]\!]$ *is the subgraph of* u*, and that the* k*-current* $T_u = -\partial\Sigma_u + S$ *is the graph of* u.

We consider the following

QUESTION: *does* $u \in BV_*(\Omega)$ *imply that* T_u *is rectifiable?*

We remark that the rectifiability of T_u is a very useful property—for instance it yields the approximate differentiability of u. We also remark that, by the Boundary Rectifiability Theorem, T_u is rectifiable if and only if $\|T_u\| < +\infty$.
The answer to the question above is

- Yes, in special situations (for instance if M is a nice open domain in $\mathbf{R}^n$),
- No, in general [ADS;Remark 2.11]

1.5. BV functions over a rectifiable current. The property of T_u being rectifiable is so important to us that we use it to build a definition.

DEFINITION 1.5.1 [ADS] *With the notation of section 1.2, let* $u : M \to \mathbf{R}$ *be* $\mathcal{H}^k$*-summable. We say that* $u \in BV(S)$ *if* $\|\Sigma_u\| < +\infty$ *and* $\|T_u\| < +\infty$.

Clearly, if $u \in BV(S)$ then T_u is rectifiable, and it's easy to see that $BV(S) \subset BV_*(S)$ (and the inclusion is proper, by what we said in 1.4).
The boundedness of $\|T_u\|$ is the crucial point in definition 1.5.1 and it is useful to have the following characterizazion.

PROPOSITION 1.5.2 [ADS] *Assume that* $\partial S = 0$ *(a similar result holds in the general case). Set*

$$\|T_u^{(1)}\| = \sup\Big\{ \int_M \sum_j \sum_\alpha \psi_j(u(x)) R^\alpha_\tau(\varphi_{j,\alpha}) d\mathcal{H}^k \quad | \\ \varphi_\alpha \in C_o^\infty(\mathbf{R}^n), \psi \in C_o^\infty(\mathbf{R}), \sum_\alpha (\sum_j \psi_j'(y)\varphi(x)_{j,\alpha})^2 \le 1 \Big\} \tag{1.5.1}$$

where $\alpha \in \Lambda(n, k-1)$*. Then one has*

$$\|T_u^{(1)}\| \le \|T_u\| \qquad \text{and} \qquad \|T_u\| \le \|S\| + \|T_u^{(1)}\|.$$

If $M = \mathbf{R}^n$ *then* $\|T_u^{(1)}\| = \|\partial(uS)\|$.

Proof. Just write the mass of T_u as the least upper bound of the evaluations of T_u on suitable test forms in $\mathbf{R}^n \times \mathbf{R}$, and then rewrite the integrals over M.

We remark the analogy between (1.5.1) and the formula (1.3.1) that defines $\|T_u^{(1)}\|$, with the notable difference that in (1.5.1) we have the non-linear function $\psi(u(x))$ in place of $u(x)$.
Both definition 1.3.1 and definition 1.5.1 can be extended to rectifiable currents with general integer multiplicity.

1.6. Compactness and closure results for $BV(S)$. Of course one is interested in having some compactness and closure results for sequences of functions u_i in the space $BV(S)$, with respect to suitable weak convergences. The situation is quite good if S is kept fixed, and is less good if S is variable.

PROPOSITION 1.6.1 [Sc] *Assume that $S = [\![M, \tau, \rho]\!]$ is an integer multiplicity rectifiable current $u_j \in BV(S)$ for $j \in \mathbf{N}$, $\|\Sigma_{u_j}\| + \|T_{u_j}\| \leq C < +\infty$ $\forall j \in \mathbf{N}$. Then there exists a subsequence u_{j_k} of u_j such that*

$$u_{j_k} \to u \qquad \text{in } L^1(\mathcal{H}^k \llcorner M)$$

for some $u \in BV(S)$.

Now we come to the case of a variable domain. Let us consider the following situation:

- S_j are k-dimensional integer multiplicity rectifiable currents in $\mathbf{R}^n$
- $u_j \in BV(S_j)$
- T_j is the graph of u_j on S_j as in 1.4
- Σ_j is the subgraph of u_j on S_j as in 1.4
- $\|T_j\| + \|\Sigma_j\| \leq C < +\infty \qquad \forall j$
- $T_j \rightharpoonup T$, $S_j \rightharpoonup S$ for $j \to \infty$

In the situation above we ask the following

QUESTION: *is T the graph of a BV function on S?*

We have some answer to this question:

- No, in general [ADS; examples 4.1 and 4.4], but
- Yes, under some additional assumptions.

For instance [ADS; theorem 4.3], the answer is yes if one assumes also

$$S_j = \partial E_j, \text{ where } E_j \text{ are Caccioppoli sets in } \mathbf{R}^n \tag{1.6.1}$$

and

$$P(E_j) \to P(E) \qquad \text{for } j \to \infty \tag{1.6.2}$$

where P(E) is the perimeter of E.
Assumption (1.6.2) is rather strong – for instance it will not be satisfied in general by a minimizing sequence for a problem where one only has a control of the perimeter. On the other hand (1.6.1) by itself is not sufficient and some assumption like (1.6.2) seems to be necessary.

2. Curvature Varifolds and currents whose tangent is in BV

Let M be a rectifiable set of dimension k in $\mathbf{R}^n$, of finite $\mathcal{H}^k$ Hausdorff measure, and let T_xM be the tangent plane to M at x. We recall that the *varifold* associated to M is the positive measure V on $\mathbf{R}^n \times G(n,k)$ defined as

$$V(\varphi) = \int_M \varphi(x, T_x) d\mathcal{H}^k \qquad \forall \varphi \in C^0_o(\mathbf{R}^n \times G(n,k))$$

where $\mathbf{R}^n \times G(n,k) = \{k - \text{planes in } \mathbf{R}^n\}$.

We recall also that a k-plane T in $\mathbf{R}^n$ may be identified with the orthogonal projection operator $P = \{P_{ij}\}_{i,j=1,\dots,n}$ onto T. Then the tangential derivatives of P_{ij} on M contain all the information about the curvature of M.
With Hutchinson [Hu] we say that V is a *curvature varifold* in $\mathbf{R}^n$ if there exists a set $\{A_{ijk}\}_{i,j,k=1}^n$ of V-summable functions on $\mathbf{R}^n \times G(n,k)$ such that

$$\begin{aligned}\int_M \Big\{ \sum_{j=1}^n P_{ij}(x) D_j\varphi(x,P(x)) + \sum_{j,k=1}^n A_{ijk}(x,P(x)) D^*_{jk}\varphi(x,P(x)) + \\ + \sum_{j=1}^n A_{jij}(x,P(x))\varphi(x,P(x)) \Big\} \rho(x)\, d\mathcal{H}^k(x) = 0\end{aligned} \tag{2.1}$$

for $i = 1,\dots,n$, and for all $\varphi \in C^1(\mathbf{R}^n \times \mathbf{R}^{n^2})$, where $D_j\varphi$ and $D^*_{jk}\varphi$ denote the partial derivatives of φ with respect to the variables x_j and P_{jk} (freezing all other variables).
Equation (2.1) defines *globally* the functions A_{ijk} as weak derivatives of P_{ij}. The tensor A_{ijk} is the generalized second fundamental form of V, and one would like to have some way to recover the $A_{ijk}(x)$ from some *local* approximate differentiation of P_{ij} at x.
It can be useful to know how to detect information about the second fundamental form of a curvature varifold from a field of tangent k-vectors. In fact one has the proposition below.

PROPOSITION 2.1. [DS] *Let $G_o(n,k)$ be the Grassmanian of the oriented k-dimensional planes in $\mathbf{R}^n$, considered as a submanifold of $\Lambda^k\mathbf{R}^n$, and let $\pi : G_o(n,k) \to G(n,k)$ be the natural double covering. Let M and V as above. V is a curvature varifold if and only if there exist:*

a measurable field $\tau(x)$ of unit simple k-vectors and

V-summable functions $B_{j\alpha} : \mathbf{R}^n \times G_o(n,k) \to \mathbf{R}$ (for $j = 1,\dots,n$ and $\alpha \in \Lambda(n,k)$),

such that

$$\begin{aligned}\int_M \Big\{ \psi(x,\tau(x)) \sum_\alpha \sum_j \frac{\partial \pi_{ij}}{\partial p_\alpha}(\tau(x)) B_{j\alpha}(x;\tau(x)) + \sum_j \pi_{ij}(\tau(x)) D_j\psi(x,\tau(x)) + \\ + \sum_\alpha B_{i\alpha}(x;\tau(x)) D^*_\alpha \psi(x,\tau(x)) \Big\} \rho(x)\, d\mathcal{H}^k(x) = 0,\end{aligned} \tag{2.2}$$

for all $i = 1,\dots,n$, and for every even function $\psi \in C^1(\mathbf{R}^n \times \Lambda^k\mathbf{R}^n)$ having compact support with respect to the x variables (even means that $\psi(x,\tau) = \psi(x,-\tau)$). From the numbers $B_{i\alpha}(x;\tau)$ one can recover the numbers $A_{ijk}(x,\pi(\tau))$.

Proposition 2.1 leads

- to introduce *oriented* curvature varifolds [DS][D] (in (2.2) one takes test functions ψ which are not necessarily even)
- to link the notion of curvature varifold to the notion of BV function on a rectifiable current.

About the second point above one has the following result

PROPOSITION 2.2. [DS] *Let V be a curvature varifold associated to a rectifiable set M, and let $\tau(x)$ be any measurable field of unit simple tangent k-vectors for M. Then:*

- $\tau \in BV_*(S)$
- *given any fixed index α there is a field $\tau^\star$ of tangent k-vectors for M, such that $|\tau^\alpha| \in BV(S^\star)$ if $S^\star = [\![M, \tau^\star, 1]\!]$*
- *τ is approximately differentiable $\mathcal{H}^k$-almost everywhere on M, and the functions $A_{ijk}(x)$ may be written in terms of $ap\nabla\tau(x)$.*

For the even closer relation between *oriented* curvature varifolds and currents whose tangent is in BV, we refer to [DS].

3. SBV functions on a rectifiable current and surfaces whose curvature is concentrated on integer-dimension sets

3.1. SBV functions on a rectifiable current. In recent works (see for instance the review [Amb1]) certain BV functions of special type have been introduced (SBV functions), which are useful to model problems where the unknown functions may have free discontinuities. SBV functions on euclidean space are defined as BV functions whose gradient has no Cantor part. Also SBV functions on a rectifiable current may be defined [O]. In this case, to detect the Cantor part of the gradient of a function u, it is convenient to look at the graph of u.
Let M, τ, S be as in section 1.2. Let $u \in BV(S)$ and let $T_u = [\![R, \xi, 1]\!]$.
We also set:

$$\begin{aligned} R^0 &= \{(x,y) \in R \quad | \quad \dim\, p(T_{x,y}R) = k\} \\ R^1 &= \{(x,y) \in R \quad | \quad \dim\, p(T_{x,y}R) = k-1\} \\ S_u &= \{x \in M \quad | \quad \mathcal{H}^1(p^{-1}(x) \cap R) > 0\} \end{aligned}$$

where $p : \mathbf{R}^n_x \times \mathbf{R}^n_y \to \mathbf{R}^n_x$ is the standard projection.
Up to sets of $\mathcal{H}^k$-measure zero one has

$$R = R^0 \cup (R^1 - p^{-1}(S_u)) \cup p^{-1}(S_u) \tag{3.1.1}$$

In the special case when $k = n$ and M is an open set in euclidean space, the Cantor part of the gradient of u lives on the set $(R^1 - p^{-1}(S_u))$. The following definition is then quite natural.

DEFINITION 3.1.1 [O] *A function $u \in BV(S)$ is said to be a SBV function if $\mathcal{H}^k((R^1 - p^{-1}(S_u))) = 0$.*

For SBV functions on a rectifiable current one has the following compactness theorem.

THEOREM 3.1.2 [O] *Let M, S as above. Assume that*

$$u_h \in SBV(S) \qquad \forall h \in \mathbf{N}$$

$$\|T_{u_h}\| + \|\Sigma_{u_h}\| + \mathcal{H}^{k-1}(S_{u_h}) + \int_M |A_h|^p d\mathcal{H}^k \leq C < +\infty \qquad \forall h \in \mathbf{N}$$

for some fixed $p > 1$, where $A_h(x)$ is the density with respect to $\mathcal{H}^k \llcorner M$ of the absolutely continuous part of the tangential gradient of u_h. Then there is a subsequence (that we still denote u_h) such that

$$\begin{aligned} &u_h \to u \quad &&\text{in } L^1(\mathcal{H}^k \llcorner M) \quad \text{for some } u \in SBV(S) \\ &A_{u_h} \rightharpoonup A_u \quad &&\text{weakly in } L^p \\ &\liminf_{n\to\infty} \mathcal{H}^{k-1}(S_{u_h}) \geq \mathcal{H}^{k-1}(S_u). && \end{aligned}$$

The proof of this result follows the pattern of the recent new proof [Amb2] of the compactnes theorem in SBV. In [O] one also has a closure result for the case of a variable S.

3.2. Surfaces whose curvature is concentrated on integer dimension singular sets. For various problems where an optimal surface is sought, for instance in approximation, segmentation of images, surface reconstruction, or shape optimization, it would be useful to have a class of generalized surfaces whose curvatures are concentrated on sets of integer dimension, and which is stable with respect to a suitable weak convergence. One possible good step in this direction is certainly the recent work [Man]. Here we describe very quickly another possible approach [AO], which is for the moment restricted to boundaries of convex sets .
Let K be a bounded convex set in $\mathbf{R}^3$, with a non empty interior. For each $x \in K$ we consider the normal cone to K at x:

$$N_x(K) = \{y \in S^2 \quad | \quad (z - x) \cdot y \leq 0 \quad \forall z \in K\}$$

and the set (the unit normal bundle of K [Zae])

$$N(K) = \{(x, y) \quad | \quad x \in \partial K \, , \, y \in N_x(K)\}$$

which is rectifiable, and carries a rectifiable current, the generalized Gauss graph of ∂K [AST].
We introduce the singular sets

$$\begin{aligned} \Sigma_1 &= \{x \in \partial K \quad | \quad \mathcal{H}^1(N_x(K)) > 0\} \\ \Sigma_0 &= \{x \in \partial K \quad | \quad \mathcal{H}^2(N_x(K)) > 0\} \end{aligned}$$

The set Σ_1 has some regularity property.

THEOREM 3.2.1 *The set Σ_1 is C^2-rectifiable, i.e. [AS] it is contained (possibly except a set of zero $\mathcal{H}^1$-measure) in the union of a countable family of class C^2 1-dimensional manifolds .*

The result above was first conjectured Giovanni Alberti, and was then proved independently, in different ways, in [Alb] and [AO].
By the C^2-rectifiability one can define (locally, not by global integration by parts formulae) a curvature $c(x)$ for $\mathcal{H}^1$-almost all $x \in \Sigma_1$. Hence one can consider the functional

$$F(K) = \int_{\partial K \setminus \Sigma_1} \{1 + |A|^2 + K^2\} d\mathcal{H}^2 + \int_{\Sigma_1 \setminus \Sigma_0} (1 + c^2) d\mathcal{H}^1 + \mathcal{H}^0(\Sigma_0)$$

where K and A are the Gauss curvature and the second fundamental form of ∂K, which are defined $\mathcal{H}^2$-almost everywhere. In the class of all convex sets the functional $F(K)$ is not very well behaved. The behaviour improves in the class of special convex sets that we are going to define below.
Consider the sets

$$N_i = \{(x,y) \in N(K) \quad | \quad \dim p(T_{(x,y)}) = i\} \qquad i = 1,2$$

DEFINITION 3.2.2 *A convex set K is said to be of special type if*

$$\mathcal{H}^2(N_i - p^{-1}(\Sigma_i)) = 0 \qquad \text{for } i = 1,2$$

Definition 3.2.2 is analogous to definition 3.1.1.
One can see that if K belongs to the class $\mathcal{A}$ of the special convex sets, then the number $\mathcal{H}^2(N(K))$ is bounded by the functional $F(K)$, hence one can consider minimum problems for $F(K)$, subject to suitable side conditions as for instance $B \subset K$ for some given set B, and one obtains minimizing sequences K_j such that both K_j and the currents $N(K_j)$ converge. It is open the

QUESTION *If $K_j \in \mathcal{A}$ is a minimizing sequence of F, and if $N(K_j) \rightharpoonup N(K)$, is K still a convex set of special type?*

References

[Alb] G. Alberti, *On the structure of singular sets of convex functions*, Calc. Var. **2** (1994), 17–27

[Amb1] L. Ambrosio, *Variational problems in SBV*, Acta Appl. Math. **17** (1989), 1–40

[Amb2] L. Ambrosio, *A new proof of the SBV compactness theorem.*, To appear in Calc.Var.

[ADS] G. Anzellotti, S. Delladio, G. Scianna, *BV Functions over Rectifiable Currents* , To appear in Annali di Matematica Pura e Applicata

[AO] G. Anzellotti, E. Ossanna, *Singular sets of convex bodies, and surfaces with generalized curvatures* , To appear in Manuscripta Math.

[AST] G. Anzellotti, R. Serapioni, I. Tamanini, *Curvatures, Functionals, Currents*, Indiana Univ. Math. J. **39** (1990), 617–669.

[D] S. Delladio, *Do generalized Gauss Graphs induce Curvature Varifolds?*, Preprint Trento UTM 453 (1995)

[DS] S. Delladio, G. Scianna, *Oriented and non Oriented Curvature Varifolds*, To appear in Proc.Royal Soc. Edinburgh

[FE] H. Federer, *Geometric Measure Theory*, Springer Verlag, Berlin-Heidelberg-New York(1969).

[HU] J.E. Hutchinson, *Second fundamental form for varifolds and the existence of surfaces minimizing curvature*, Indiana Univ. Math. J. **35** (1986), 45–71.

[O] E. Ossanna, *SBV functions over a rectifiable current and a compactness theorem*, Preprint Trento 1995

Progress in Nonlinear Differential Equations
and Their Applications, Vol. 25

Remarks on a Numerical Study of Convexity, Quasiconvexity, and Rank One Convexity*

BERNARD DACOROGNA* AND JEAN-PIERRE HAEBERLY⋆

∗ Département de Mathématiques, Ecole Polytechnique Fédérale de Lausanne, 1015 Lausanne, Switzerland

⋆ Department of Mathematics, Fordham University, Bronx, NY 10458-5165, U.S.A.

1. Introduction

In the calculus of variations, the notions of convexity, quasiconvexity, and rank one convexity play an important role. In this paper we will discuss some numerical results concerning these notions.

We begin by recalling some definitions. Let $\mathbf{R}^{2\times 2}$ denote the set of 2×2 real matrices. Let $f : \mathbf{R}^{2\times 2} \longrightarrow \mathbf{R}$ be continous.

DEFINITION *The function f is* rank one convex *if*

$$f(tA + (1-t)B) \leq tf(A) + (1-t)f(B) \tag{1.1}$$

for every $t \in [0,1]$, and every $A, B \in \mathbf{R}^{2\times 2}$ with $\det(A - B) = 0$, where $\det A$ denotes the determinant of the matrix A.

DEFINITION *The function f is* quasiconvex *if*

$$f(A) \leq \frac{1}{\text{meas}\,\Omega} \int_{\Omega} f(A + \nabla\varphi(x))\,dx \tag{1.2}$$

for every bounded open set $\Omega \subset \mathbf{R}^2$, every $A \in \mathbf{R}^{2\times 2}$, and every $\varphi \in C_0^\infty(\Omega; \mathbf{R}^2)$.

It is well known that 1.2 implies 1.1, while, obviously, the convexity of f implies 1.2. In concrete examples it might be very hard to decide whether a given function is quasiconvex or not. In particular, it is a long standing question to know if 1.1 and 1.2 are equivalent. (In higher dimensions they are not equivalent, see Sverak [13].)

We propose some numerical algorithms to determine the convexity, quasiconvexity, and rank one convexity of a function. In particular, we study two families

*This research was supported in part by a grant from the *Fonds National Suisse pour la Recherche Scientifique*.

of examples, namely

$$f(A) = |A|^{2\alpha}\left(|A|^2 - \gamma \det A\right) \tag{1.3}$$

$$f(A) = |A|^{4\alpha} - 2^{2\alpha-1}\gamma\left(\left(\det A\right)^2\right)^{\alpha} \tag{1.4}$$

where $\alpha \geq 0$, $\gamma \in \mathbf{R}$, and $|A|$ denotes the Euclidean norm of a matrix (i.e. $|A|^2 = \sum_{i,j=1}^{2} A_{ij}^2$).

These families of examples have been studied analytically for the rank one convexity by Dacorogna, Douchet, Gangbo, and Rappaz [9]. For the particular case $\alpha = 1$ in the first example, the convexity, quasiconvexity, and rank one convexity were analyzed by Dacorogna and Marcellini [7], and by Alibert and Dacorogna [1]. The case $\alpha < 1$ in the second example has been studied by Ball and Murat [2].

Concerning numerical computations on these examples and on related questions see Dacorogna, Douchet, Gangbo, and Rappaz [9], Gremaud [10], Brighi and Chipot [3], Collins, Kinderlehrer, and Luskin [5], Collins and Luskin [6], Chipot and Lécuyer [4].

Our numerical results are in accordance with the analytical ones whenever these are known. They also tend to show that for the families of examples studied, quasiconvexity and rank one convexity are equivalent. In a forthcoming paper, we will present the numerical analysis of other families of functions, as well as examples in higher dimensions. We will also discuss another notion of convexity, namely polyconvexity, that is relevant in the calculus of variations (see [8] for a precise definition).

2. Analytical Results

We briefly recall the known results on the convexity properties of the two families of examples mentioned above (see [9] for details).

Example 1: Let $\alpha \geq 1$ and define

$$\gamma_1 = \left(1 + \frac{1}{\alpha}\right) \min_{t>0} \left\{ \frac{t^4 + 2(\alpha+1)t^2 + 2\alpha + 1}{3t^3 + (2\alpha+1)t} \right\}$$

and

$$\gamma_2 = 1 + \sqrt{1 - \frac{1}{2\alpha} - \frac{1}{2\alpha^2}}$$

The function

$$f_\gamma(A) = |A|^{2\alpha}\left(|A|^2 - \gamma \det A\right)$$

is rank one convex if and only if $|\gamma| \leq \gamma_{\mathrm{r}}$, where

$$\gamma_{\mathrm{r}} = \begin{cases} \gamma_1 & \text{if } \alpha \in \left[1, \frac{9+5\sqrt{5}}{4}\right), \\ \gamma_2 & \text{if } \alpha \geq \frac{9+5\sqrt{5}}{4} \end{cases}$$

Also, there exist constants γ_{c} and γ_{q} such that f_γ is convex, respectively quasiconvex, if and only if $|\gamma|$ is less than or equal to γ_{c}, respectively γ_{q}.

When $\alpha = 1$, the value of γ_{c} is known to be equal to $\frac{4}{3}\sqrt{2}$, while $\gamma_{\mathrm{r}} = \frac{4}{\sqrt{3}}$ in this case. Furthermore, there is an $\epsilon > 0$ such that $\gamma_{\mathrm{q}} = 2 + \epsilon$, but the value of ϵ is not known. When $\alpha > 1$ the values of γ_{c} and γ_{q} have not been determined.

Example 2: Let $\alpha \geq \frac{1}{4}$ and define

$$\gamma_1 = \frac{\frac{2(2\alpha-1)}{4\alpha-1}}{\left(1 - \frac{\alpha}{(2\alpha-1)^2}\right)^{2(\alpha-1)}}$$

and

$$\gamma_2 = \frac{2(2\alpha + (2\alpha - 1)z)}{(2\alpha - 1)(1 - z)(1 - z^2)^{\alpha-1}}$$

where

$$z = \frac{-(4\alpha^2 - 1) + \left[(4\alpha^2 - 1)^2 - 8(2\alpha - 1)(\alpha - 1)(4\alpha - 1)\right]^{1/2}}{4(2\alpha - 1)(\alpha - 1)}$$

The function

$$f_\gamma(A) = |A|^{4\alpha} - 2^{2\alpha-1}\gamma\left((\det A)^2\right)^\alpha$$

is rank one convex if and only if $\gamma \leq \gamma_{\mathrm{r}}$, where

$$\gamma_{\mathrm{r}} = \begin{cases} 0 & \text{if } \frac{1}{4} \leq \alpha < 1 \\ 1 & \text{if } \alpha = 1 \\ \gamma_2 & \text{if } 1 < \alpha < 1 + \frac{1}{\sqrt{2}} \\ \gamma_1 & \text{if } \alpha \geq 1 + \frac{1}{\sqrt{2}} \end{cases}$$

When $\alpha \in [\frac{1}{4}, 1]$, f_γ is convex if and only if it is rank one convex.

In general, there exist γ_{q}, $\gamma_{\mathrm{c}}^+ \geq 0$, and $\gamma_{\mathrm{c}}^- \leq 0$, such that f_γ is quasiconvex if and only if $\gamma \leq \gamma_{\mathrm{q}}$, and is convex if and only if $\gamma \in [\gamma_{\mathrm{c}}^-, \gamma_{\mathrm{c}}^+]$. But for $\alpha > 1$, the values of γ_{q}, γ_{c}^+, and γ_{c}^-, are not known.

Observe that f_γ is homogeneous of degree $2(\alpha + 1)$ in example 1 and homogeneous of degree 4α in example 2. Also, f_γ is always affine in γ, i.e. it can be written as

$$f_\gamma(A) = f_1(A) + \gamma f_2(A)$$

for some functions f_1 and f_2 independent of γ.

3. Numerical Methods

We first discuss numerical methods for computing γ_{c} (γ_{c}^+) and γ_{r}.

Given a function $f : \mathbf{R}^{2\times 2} \longrightarrow \mathbf{R}$, it is easy to show that it is rank one convex if and only if the function

$$\sigma(t) = f(A + tu \otimes v)$$

is convex in t for every $A \in \mathbf{R}^{2\times 2}$ and for every $u, v \in \mathbf{R}^2$. Here $u \otimes v$ is the rank one matrix obtained from the vectors $u = (u_1, u_2)$, $v = (v_1, v_2)$ as follows:

$$u \otimes v = \begin{pmatrix} u_1 v_1 & u_1 v_2 \\ u_2 v_1 & u_2 v_2 \end{pmatrix}$$

Let us assume that f is C^1, and let us define a function

$$h : \mathbf{R}^{2\times 2} \times \mathbf{R}^{2\times 2} \longrightarrow \mathbf{R}$$

as follows:

$$h(A, B) = f(A + B) - f(A) - \langle \nabla f(A), B \rangle.$$

Here $\langle\ ;\ \rangle$ denotes the scalar product on $\mathbf{R}^{2\times 2}$. It is well-known that f fails to be convex if and only if

$$\inf_{A,B} h(A, B) < 0 \tag{3.5}$$

while it fails to be rank one convex if and only if

$$\inf_{A,u,v} h(A, u \otimes v) < 0. \tag{3.6}$$

Now let $f = f_\gamma = f_1 + \gamma f_2$ be a function from example 1 or 2, and let us write h_γ for h. We have

$$h_\gamma(A, B) = h_1(A, B) + \gamma h_2(A, B)$$

where, in example 1,

$$\begin{aligned} h_1(A, B) &= |A + B|^{2(\alpha+1)} - |A|^{2(\alpha+1)} - 2(\alpha + 1)\langle A, B\rangle |A|^{2\alpha} \\ h_2(A, B) &= |A|^{2(\alpha-1)} \left[|A|^2 \langle \tilde{A}, B\rangle + |A|^2 (\det A) + 2\alpha (\det A) \langle A, B\rangle \right] \\ &\quad - |A + B|^{2\alpha} \det(A + B) \end{aligned}$$

while in example 2,

$$\begin{aligned} h_1(A, B) &= |A + B|^{4\alpha} - |A|^{4\alpha} - 4\alpha \langle A, B\rangle |A|^{4\alpha-2} \\ h_2(A, B) &= 2^{2\alpha-1} \left[2\alpha (\det A)^{2\alpha-1} \langle \tilde{A}, B\rangle + (\det A)^{2\alpha} - (\det(A + B))^{2\alpha} \right]. \end{aligned}$$

Here $\tilde{A}$ is the gradient of the determinant function evaluated at A, namely

$$\tilde{A} = \begin{pmatrix} A_{22} & -A_{21} \\ -A_{12} & A_{11} \end{pmatrix}$$

To compute γ_c (or γ_c^+) we may proceed as follows.

ALGORITHM 1. *Choose an initial value of γ large enough that f_γ is not convex.*

2. *Use a minimization algorithm to compute a pair of matrices (A, B) for which $h_\gamma(A, B) < 0$.*

3. Set γ equal to $-\frac{h_1(A,B)}{h_2(A,B)}$.

4. Go to step 2.

The algorithm stops when the minimization procedure fails to produce (A, B) with $h_\gamma(A, B) < 0$. The current value of γ is then the computed approximation to γ_c.

An identical appoach can be used to compute γ_r but with $h_1(A, B), h_2(A, B)$ replaced with $h_1(A, u \otimes v), h_2(A, u \otimes v)$.

This method works well and is perfectly suitable to compute γ_c or γ_r, where the optimization takes place in a space of very small dimension. The method is inherently inefficient, however. Indeed, the minimization procedure needs to be applied to the function h_γ for a large number of values of γ. This amounts to a considerable computational effort, even if one uses the computed pair (A, B) of the previous step as the starting point of the minimization process of the current step. In the case of quasiconvexity, where the optimization takes place in a space of very large dimension, it is thus much preferable to compute the threshold values γ_q directly as the minima of certain functions. In order to describe how to construct these functions, we observe the following. Suppose that we are given a family of function $\phi_\gamma : \mathbf{R}^n \longrightarrow \mathbf{R}$ of the form

$$\phi_\gamma(x) = \phi_1(x) + \gamma\phi_2(x)$$

where ϕ_1, ϕ_2 are independent of γ.

If $\phi_1(x) \geq 0$ for all $x \in \mathbf{R}^n$, then

$$\begin{aligned}
\phi_\gamma(x) \geq 0 \text{ for all } x &\iff \phi_1(x) \geq -\gamma\phi_2(x) \text{ for all } x \\
&\iff \begin{cases} -\dfrac{\phi_1(x)}{\phi_2(x)} \geq \gamma \text{ for all } x \text{ with } \phi_2(x) < 0 \\[2ex] -\dfrac{\phi_1(x)}{\phi_2(x)} \leq \gamma \text{ for all } x \text{ with } \phi_2(x) > 0 \end{cases} \\
&\iff \gamma^- \equiv \sup_{\phi_2(x)>0} -\frac{\phi_1(x)}{\phi_2(x)} \leq \gamma \leq \gamma^+ \equiv \inf_{\phi_2(x)<0} -\frac{\phi_1(x)}{\phi_2(x)}.
\end{aligned}$$

By assumption, $0 \in [\gamma^-, \gamma^+]$. Hence

$$\min\{\gamma^+, -\gamma^-\} = \inf_x \frac{\phi_1(x)}{|\phi_2(x)|}.$$

In particular, if $\gamma^+ = -\gamma^- = \gamma$, then

$$\gamma = \inf_x \frac{\phi_1(x)}{|\phi_2(x)|}. \tag{3.7}$$

We now show how 3.7 can be used to estimate γ_q in examples 1 and 2. Recall that in the definition of quasiconvexity it is sufficient to take $\Omega = [0,1] \times [0,1]$, the unit square in $\mathbf{R}^2$. Thus the function $f_\gamma : \mathbf{R}^{2\times 2} \longrightarrow \mathbf{R}$ is quasiconvex if and only if

$$\int_\Omega f_\gamma(A + \nabla\varphi(x_1, x_2))\, dx_1 dx_2 \; - \; f_\gamma(A) \geq 0 \tag{3.8}$$

for all $A \in \mathbf{R}^{2\times 2}$ and all $\varphi \in C_0^\infty(\Omega;\mathbf{R}^2)$, the space of smooth functions from Ω to $\mathbf{R}^2$ with compact support in Ω. Recall further that it is possible to use Lipshitz functions that are periodic on the boundary instead. In order to approximate the left hand side of 3.8, we use finite elements. More precisely, we fix an integer $m \geq 2$, and we construct a regular mesh of triangles of type 1 on Ω as follows. We partition the interval $[0,1]$ into m subintervals of equal length, and then divide each square of the resulting grid into two triangles, for a total of $2m^2$ triangles. We then replace the function φ with its values at the $(m-1)^2$ interior nodes in the case of zero boundary conditions, or at the m^2 independent nodes in the case of periodic boundary conditions. Our approximation is then a function

$$k_{\gamma,m} : \mathbf{R}^n \longrightarrow \mathbf{R}$$

where n is either $(m-1)^2+4$ or m^2+4 depending on the choice of boundary conditions. Note that $k_{\gamma,m}$ is also affine in γ. As before, we write $k_{\gamma,m}(x) = k_{1,m}(x) + \gamma k_{2,m}(x)$. Note further that $k_{1,m}(x) \geq 0$ for all $x \in \mathbf{R}^n$ since f_1 is a convex function in examples 1 and 2. Now let

$$\gamma_{q,m} = \inf_x \frac{k_{1,m}(x)}{|k_{2,m}(x)|}.$$

By approximation, we have the following:

$$\gamma_{\mathrm{q}} = \lim_{m\longrightarrow\infty} \gamma_{q,m}.$$

For the minimization, we have used an extension of the conjugate gradient algorithm of Polak and Ribière due to Shanno [11], and implemented in the subroutine CONMIN developed by Shanno and Phua [12]. The algorithm uses a cubic interpolation for the line search.

4. Numerical results: example 1

The algorithms were programmed in FORTRAN, and all computations were done on Silicon Graphics workstations equipped with R4000 and R8000 processors.

We have computed γ_{c}, γ_{r}, γ_{q}, for α between 1 and 8 in steps of 0.01. For the case of convexity and rank one convexity, the minimization routine was initialized with a random starting point for every value of α. For the case of quasiconvexity, the minimization algorithm was initialized with a random starting point for $\alpha = 1$. Henceafter, the starting point was taken to be the computed solution at the previous value of α.

When $\alpha = 1$, so that the correct values of γ_{c} and γ_{r} are known, the results were as follows:

	computed value	correct value	error
γ_{c}	1.88561822236594	1.88561808316413	$\sim 1.4\times 10^{-7}$
γ_{r}	2.30941239674591	2.3094010767585	$\sim 1.2\times 10^{-5}$
γ_{q}	2.30940590626079		

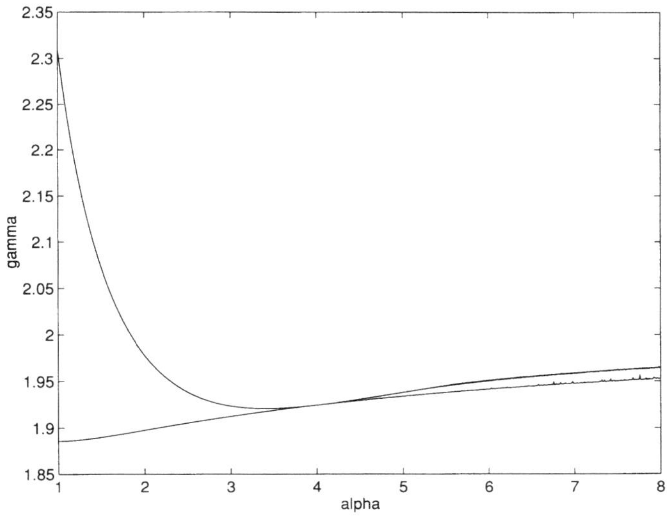

Figure 1: Graphs of γ vs α for convexity, quasiconvexity, and rank one convexity for example 1 (note that the curves for rank one convexity and quasiconvexity are, here, indistinguishable).

More generally, the worst error between the computed values of γ_{r} and the exact values, when known, was about 1.4×10^{-3}, while the largest gap between the computed values of γ_{r} and γ_{q} was about 9.2×10^{-4}.

The computed values of γ_{c}, γ_{r}, γ_{q} are plotted versus α in figure 1. We used periodic boundary conditions. The meshsize for the computation of γ_{q} was $m = 20$, resulting in 404 variables. Observe that the curve for γ_{q} and γ_{r} are nearly identical, and are thus indistinguishable on the plot.

Figure 2 exhibits an interesting phenomenon. The data is the same as that in figure 1 except that the minimization routine for the case of quasiconvexity is initialized with a random starting point for every value of α. As before we observe very good agreement between the γ_{r} and γ_{q} curves for $1 \leq \alpha \leq 4.5$. As α passes through the value at which the rank one convexity and the convexity curves become tangent, the computation of γ_{q} encounters some serious difficulties. This is due to the appearance of local minima whose basin of attraction grows progressively larger as α increases. Eventually, the algorithm, when started at a random point, can no longer compute a meaningful γ_{q}. An attempt to remedy this deficiency was proposed in [10], replacing the deterministic minimization algorithm with a stochastic one based on the method of simulated annealing. The computational cost is enormous, however, and the results obtained are no better than those displayed in figure 1.

Finally, in figures 4 and 5 we illustrate the differences between the two choices of boundary conditions. Figure 4 displays one component of the solution for the computed γ_{q} in example 1 with $\alpha = 1$ when one imposes zero boundary conditions,

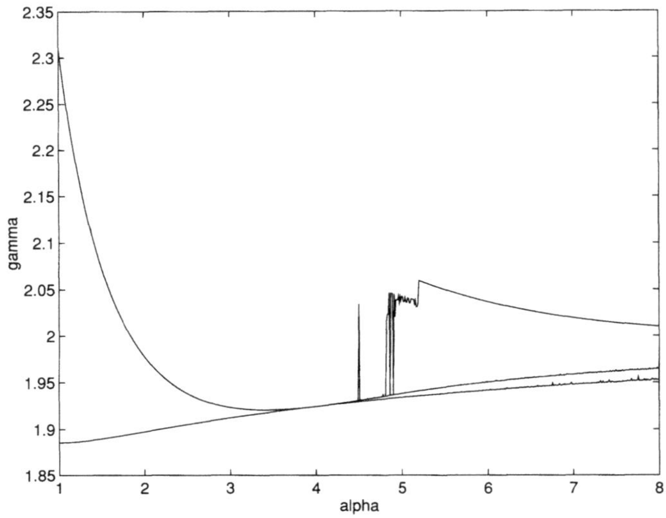

Figure 2: Graphs of γ vs α for convexity, quasiconvexity, and rank one convexity for example 1 with random starting point. The bottom curve is γ_c, the next highest γ_r, and the top one γ_q.

while figure 5 shows the solution with periodic boundary conditions. Each graph is displayed as viewed from two different view angles to emphasize the oscillatory structure. Here $m = 100$ resulting in 9805 nad 10004 variables respectively.

5. Numerical results: example 2

The results of the computations in the case of example 2 are very similar to those obtained for example 1. We computed γ_c, γ_r, and γ_q for α between 1 and 3 in increment of 0.01, using the same procedure as that used to produce figure 1. These values are plotted against α in figure 3. We took $m = 50$ and periodic boundary conditions, resulting in 2504 variables. The worst error between the computed value of γ_r and the correct values was about 1.8×10^{-3}, while the largest gap between the computed values of γ_r and γ_q was about 1.4×10^{-3}. When $\alpha = 1$, the computed value of γ_q was 1.00000808500197, yielding an error of order 10^{-6}. The gap between the computed γ_r and γ_q increases as α approaches 3.

Overall, example 2 appears slightly less amenable to numerical investigation than example 1. Local minima cause difficulties for all values of α in the chosen interval [1,3]. Furthermore, with $m = 20$, the gap between the γ_r and γ_q curves becomes apparent on the plot as α approaches 2. This gap can be very effectively controlled by the choice of m, however, yielding the plot of figure 3 for $m = 50$.

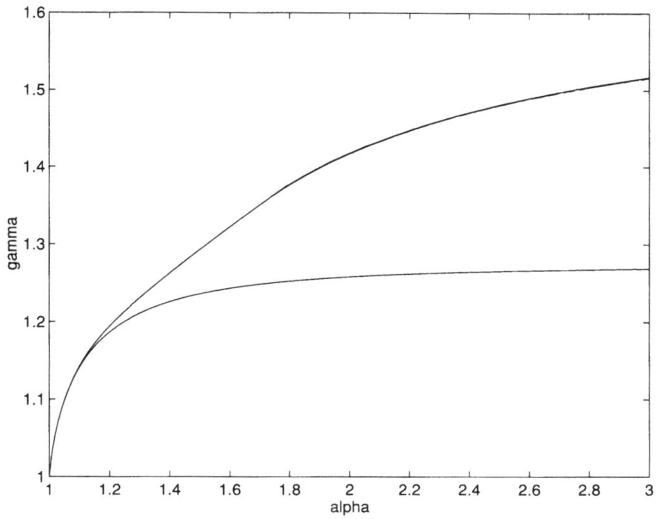

Figure 3: Graphs of γ vs α for convexity, quasiconvexity, and rank one convexity for example 2 (again the curves for rank one convexity and quasiconvexity are, here, indistinguishable).

6. Conclusions

We conclude with a few remarks about the behavior of the numerical methods and the quality of the numerical results.

- Agreement between the computed values and the correct values, when known, are very good.
- When the gap between the computed values of γ_q and γ_r is small for a certain value of m, increasing m further yields no significant gain. When it is large, it can be successfully reduced by increasing m.
- Using periodic boundary conditions gives slightly better values of γ_q than using zero boundary conditions.

We would like to point out that, to the best of our knowledge, this work is the first systematic numerical investigation of convexity, rank one convexity, and quasiconvexity. In prior work, only γ_q has been computed in a few cases. We have computed γ_c, γ_r, and γ_q, for about a thousand different functions. Moreover, the results we obtain are sharper than older results we are aware of.

It is currently not known whether a function

$$f : \mathbf{R}^{2\times 2} \longrightarrow \mathbf{R}$$

is rank one convex if and only if it is quasiconvex. The computations presented above provide some evidence that, for the families of functions in examples 1 and 2, this is indeed the case.

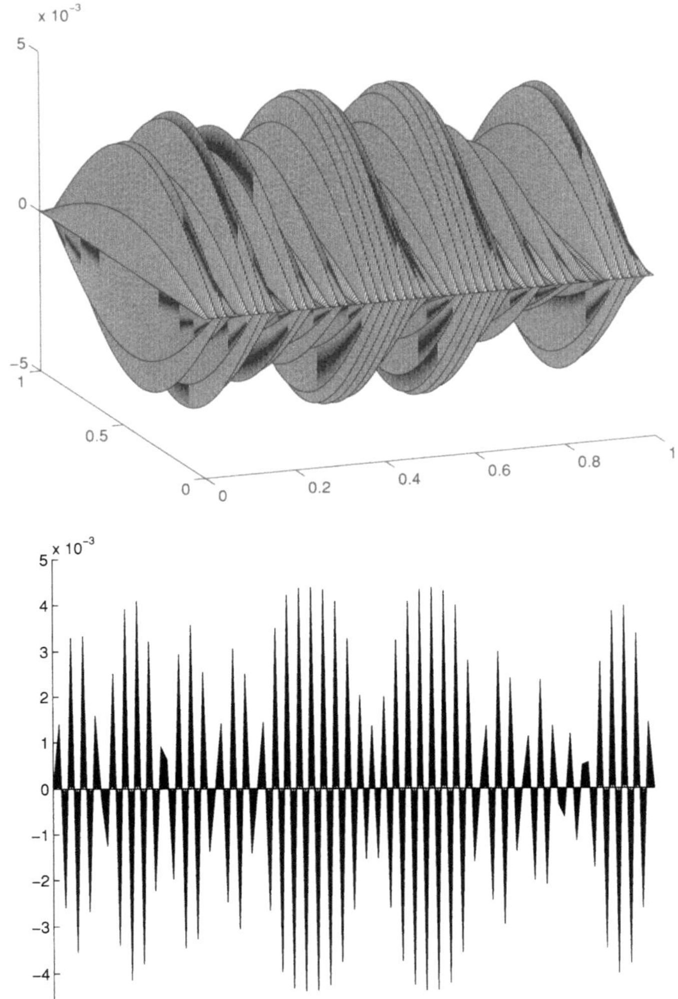

Figure 4: Graph of one component of the solution with 0 boundary conditions for quasiconvexity in example 1 with $\alpha = 1$, viewed from two different angles. The bottom figure illustrates clearly the oscillatory structure of the function. Here $m = 100$ and the computed value of γ_q is 2.3095955016.

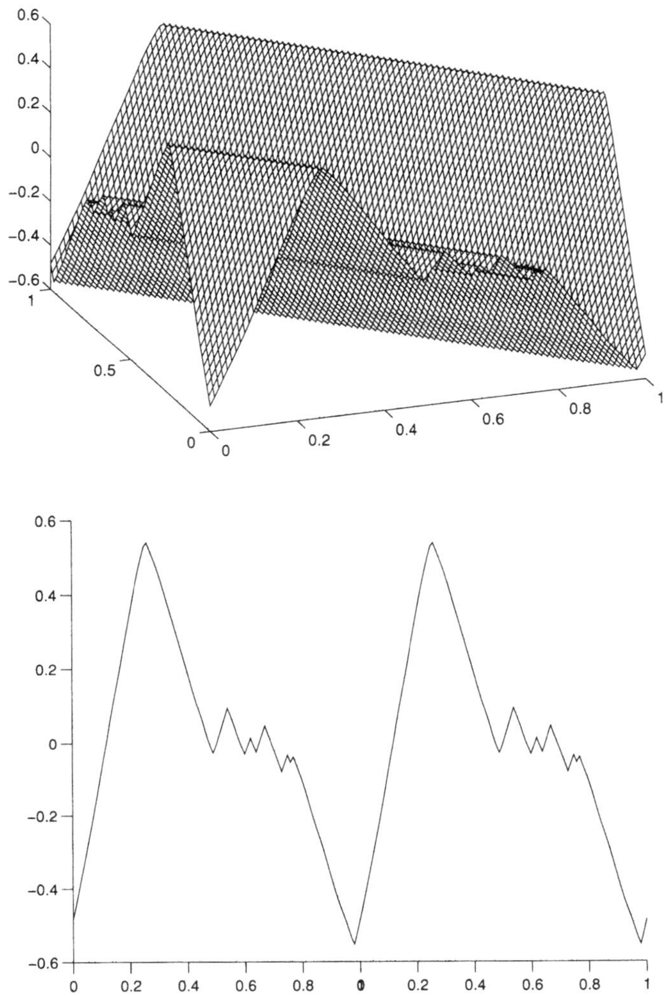

Figure 5: Graph of one component of the solution with periodic boundary conditions for quasiconvexity in example 1 with $\alpha = 1$, viewed from two different angles. The bottom figure illustrates clearly the oscillatory structure of the function. Here $m = 100$ and the computed value of γ_q is 2.3094670086.

References

[1] J.J. Alibert and B. Dacorogna, *An example of a quasiconvex function that is not polyconvex in two dimensions*, Arch. Rational Mech. Anal. 117 (1992), pp. 155–166.

[2] J.M. Ball and F. Murat, $W^{1,p}$ *quasiconvexity and variational problems for multiple integrals*, J. Funct. Anal. 58 (1984), pp. 225–253.

[3] B. Brighi and M. Chipot, *Approximated Convex Envelope of a Function*, SIAM J. Numer. Anal. 31 (1994), pp. 128–148.

[4] M. Chipot and V. Lécuyer, *Analysis and computations in the four-well problem*, Preprint.

[5] C. Collins, D. Kinderlehrer, and M. Luskin, *Numerical approximation of the solution of a variational problem with a double well potential*, SIAM J. Numer. Anal. 28 (1991), pp. 321–332.

[6] C. Collins and M. Luskin, *Optimal order error estimates for the finite element approximation of the solution of a nonconvex variational problem*, Math. Comp. 57 (1991), pp. 621–637.

[7] B. Dacorogna and P. Marcellini, *A counterexample in the vectorial calculus of variations*, in *Material instabilities in continuum mechanics*, proceedings edited by J.M. Ball, Oxford Science Publc., Oxford (1988), pp. 77–83.

[8] B. Dacorogna, *Direct Methods in the Calculus of Variations*, Springer, Berlin, 1989.

[9] B. Dacorogna, J. Douchet, W. Gangbo, and J. Rappaz, *Some examples of rank one convex functions in dimension two*, Proc. of Royal Soc. Edinburgh 114A (1990), pp. 135–150.

[10] P.A. Gremaud, *Numerical optimization and quasiconvexity*. To appear.

[11] D.F. Shanno, *Conjugate gradient methods with inexact searches*, Math. Oper. Res. 3 (1978), pp. 244–256.

[12] D.F. Shanno and K.H. Phua, *Remark on algorithm 500: minimization of unconstrained multivariate functions*, ACM Trans. on Math. Software 6, pp. 618–622.

[13] V. Sverak, *Rank one convexity does not imply quasiconvexity*, Proc. of Royal Soc. Edinburgh, 120A (1992), pp. 185–189.

Progress in Nonlinear Differential Equations
and Their Applications, Vol. 25

Homogeneous Fractal Spaces

UMBERTO MOSCO AND LINO NOTARANTONIO

Dipartimento di Matematica, Università La Sapienza
Piazzale Aldo Moro 2, 00185 Roma, Italy

1. Introduction

Diffusion processes on special classes of fractal sets, and related self-adjoint generators, have been studied in recent years by several authors, both by probabilistic and analytic methods. In particular, for the so-called Sierpinski gasket the construction of the "Brownian motion" and "Laplace" operator has been done by [Ku1], [BP], [Ki], [FS].

One important aspect of this theory consists in the extension of the classic Weyl's spectral asymptotic formula. If $-\Delta$ is the classic Laplace operator in a bounded open domain of the Euclidean space $\mathbb{R}^D$, with Dirichlet boundary condition, then the number $\rho(\lambda)$ of the eigenvalues (counted with their multiplicity) that do not exceed λ increases with rate $\lambda^{D/2}$ as $\lambda \to \infty$. If we consider now the Laplace operator $-\Delta$ on the (finite) Sierpinski gasket in $\mathbb{R}^D$, then $\rho(\lambda)$ increases with rate $\lambda^{D_s/2}$, with the exponent

$$D_s = 2\left(\frac{\log N}{\log(N+2)}\right) ,$$

where $N = D+1$. This spectral behaviour was first shown by physical arguments by Rammal and Toulouse in [RT], who called D_s the spectral exponent. The mathematical result follows from the estimates given in [BP] for the transition probability density of the Brownian motion on the gasket. It was also proved by a Dirichlet form approach in [FS].

What is remarkable about the spectral behaviour of the Sierpinski gasket is that the spectral exponent D_s differs from the Hausdorff (fractal) dimension of the gasket, $D_f = \left(\frac{\log N}{\log 2}\right)$, even if quite naturally D_f could be conjectured to take the place of the euclidean dimension D of the classic case (and such a conjecture had indeed been done before Rammal and Toulouse's result).

A natural question arises, namely, if the spectral dimension, which is of variational nature, can be given a "geometric" characterization, in analogy with the classic case. Our aim in this paper is to provide such a characterization, by relying on the notion of *space of homogeneous type*, which occurs in harmonic analysis, see [CW] and [S], as a generalization of the notion of homogeneous group.

A special case of space of homogeneous type is a quasi-metric space X on which a Borel measure μ exists, such that for some constants $v > 0$, $c_0 > 0$, we have $0 < c_0\mu(B(x,R))\left(\frac{r}{R}\right)^v \le \mu(B(x,r))$, $0 < r \le R$. If the opposite inequality also holds for some constant $c' > 0$, then we say that v is the *homogeneous dimension* of X.

By saying that X is a *quasi-metric* space, we mean that X is a topological space on which a quasi-distance d — that is a function $d : X \times X \to \mathbb{R}_+$ that verifies the axioms of a metric with the triangle inequality satisfied up to a multiplicative constant $c_T > 0$ — is defined, such that the quasi-balls of d are a base of open neighborhoods for the given topology of X.

We will show below that a quasi-distance d can be defined on the Sierpinski gasket K, that gives K the structure of a space of homogeneous type. Then, the spectral dimension D_s is seen to coincide with the homogeneous dimension v of K.

Moreover, the functions of finite energy on K, that is, the function in the domain of the Dirichlet form associated with the Laplace operator on K, are Hölder continuous with respect to d, with Hölder exponent $\beta = 1 - \left(\frac{v}{2}\right)$. Since $v < 2$, this corresponds to a Morrey's type imbedding, as given in [BM1]. As we shall see, this is in agreement, and also explains, previous direct calculations due to [BP] and [Ko].

Both results suggest that the homogeneous dimension introduced in this paper is a good candidate to the role of intrinsic dimension on the gasket and, possibly, on more general families of fractals, [L], [Ku2], [F].

2. The Sierpinski gasket

Let $\mathbb{R}^D$ be the D-dimensional Euclidean space, $D \geq 2$, and let $|x - y|$ be the Euclidean distance of $x, y \in \mathbb{R}^D$. Let $N = D + 1$. For every $p_i \in \mathbb{R}^D$, $i = 1, \ldots, N$, such that the vectors $p_1 p_j$, $j = 2, \ldots, N$, are independent, we denote the D-dimensional simplex of vertices $p_1, \ldots, p_N$ by $M = |p_1, \ldots, p_N|$, its vertices by $V(M) = \{p_1, \ldots, p_N\}$, the midpoints of edges by $\frac{1}{2} p_i p_j$, $i \neq j$, with additional notation $\frac{1}{2} p_i p_i = p_i$, and we introduce N new simplices $M_i := \left|\frac{1}{2} p_1 p_i \frac{1}{2} p_2 p_i \cdots \frac{1}{2} p_N p_i\right|$, $i = 1, \ldots, N$.

The unit Sierpinski gasket in $\mathbb{R}^D$ is constructed as follows. Let K_0 be a D-dimensional simplex with all edges of unit length. We put $F_0 = \{K_0\}$ and for $m \geq 1$ we define inductively

$$F_m = \bigcup_{M \in F_{m-1}} \{M_i : 1, \ldots, N\} .$$

We consider the vertices $V_0 = V(K_0)$, $V_m = \bigcup_{M \in F_m} V(M)$, $m \geq 1$, and we put $V_* = \bigcup_{m \geq 0} V_m$. Then, the Sierpinski gasket K is defined to be the closure of V_* in $\mathbb{R}^D$, $K = \overline{V_*}$.

The gasket K has Hausdorff dimension, also called *fractal dimension,*

$$D_f = \frac{\log N}{\log 2} ,$$

moreover, $0 < H^{D_f}(K) < \infty$, where H^{D_f} denotes the D_f-dimensional Hausdorff measure in $\mathbb{R}^D$. The normalized measure

$$\mu = \left(H^{D_f}(K)\right)^{-1} H^{D_f} \angle K$$

is the invariant measure on K, that is, $\mu(\cdot) = \sum_{i=1}^{N} N^{-1}\mu\big(S_i^{-1}(\cdot)\big)$, where for each $i = 1, \ldots, N$, S_i is the similitude $S_i(x) = p_i + \frac{1}{2}(x - p_i)$, see [H].

3. The standard Laplacian

Following [FS], for every function $f : V_* \to \mathbb{R}$, we define

$$E^{(m)}(f,f) = \frac{2}{N}\left(\frac{N+2}{N}\right)^m \sum_{M\in F_m} \sum_{p,q\in V(M)} \big(f(p) - f(q)\big)^2 ,$$

for every $m \geq 0$. The sequence $E^{(m)}(f,f)$ is non-decreasing as $m \to \infty$. We then define

$$E(f,f) = \lim_{m\to\infty} E^{(m)}(f,f) ,$$

with domain

$$\mathcal{F} = \big\{f : V_* \to \mathbb{R}, E(f,f) < \infty\big\} .$$

It can be proved that every function $f \in \mathcal{F}$ can be uniquely extended to be an element of $C(K)$, the space of continuous functions on K, see [FS] Prop. 4.2, therefore $\mathcal{F}$ can be identified with a subset of $C(K)$. Moreover, the form E obtained by polarization, with domain $\mathcal{F}$, is a local, regular Dirichlet form in the Hilbert space $L^2(K,\mu)$, [FS], Th. 4.1. The self-adjoint operator Δ associated with the restriction of the form E to the domain $\mathcal{F}_0 = \big(f \in \mathcal{F} : f = 0 \text{on } V_0\big)$ is the *standard Laplacian* on K, with Dirichlet condition on $\partial K := V_0$. This is indeed the Friedrichs extension of the Laplacian originally defined by Kigami in [Ki], see [FS], Th. 4.2.

In dimension $D = 2$, it has been shown by [Ko], by direct calculations, that every function $f \in \mathcal{F}$ is Hölder continuous on K, with

$$|f(x) - f(y)| \leq c|x-y|^{\beta_\varepsilon} ,$$

where $\beta_\varepsilon = \frac{\log\left(\frac{5}{3}\right)}{\log 4}$, see also [BP].

The spectrum of the standard Laplacian $-\Delta$ consists only of positive eigenvalues with finite multiplicity, accumulating at $+\infty$. For every $\lambda > 0$, let us denote by $\rho(\lambda)$ the number of eigenvalues of $-\Delta$, repeated according to their multiplicity, that are not larger than λ. Then, it is proved in [FS], Th. 5.2, that

$$0 < \liminf_{\lambda\to\infty} \frac{\rho(\lambda)}{\lambda^{D_s/2}} < \limsup_{\lambda\to\infty} \frac{\rho(\lambda)}{\lambda^{D_s/2}} < \infty ,$$

where

$$D_s = 2\left(\frac{\log N}{\log(N+2)}\right) .$$

Notice that, differently than in the classic case, the ratio $\frac{\rho(\lambda)}{\lambda^{D_s/2}}$ does not admit an asymptotic limit value as $\lambda \to \infty$, indeed it oscillates sharply.

4. The homogeneous structure

We now introduce a quasi-metric on K that gives K the structure of a homogeneous space, as mentioned in the introduction. Let

$$\delta = \frac{\log(N+2)}{\log 4} .$$

We define the function $d : K \times K \to \mathbb{R}_+$, by

$$d(x,y) = |x-y|^\delta , \quad x, y \in K .$$

This is a quasi-distance on K, the triangular inequality being satisfied with the multiplicative constant $c_T = 2^{\delta-1}$. The quasi-balls of d are $B(x,R) = \{y \in K : d(x,y) < R\}$, $x \in K$, $R > 0$.

THEOREM 1 *Let $D = 2$. With the quasi-distance d and the invariant measure μ, K is a homogeneous space of dimension*

$$v = \frac{D_f}{\delta} ,$$

namely, we have

$$c^{-1}\mu(B(x,R))\left(\frac{r}{R}\right)^v \le \mu(B(x,r)) \le c\mu(B(x,R))\left(\frac{r}{R}\right)^v ,$$

for every $x \in K$, $0 < r \le R \le 1$, where $c > 0$ is a suitable constant.

We omit the proof, which can be obtained by describing the local structure of K in every neighborhood of size 2^{-m} and by exploiting the invariance property of the measure μ. A similar result also holds in higher dimensions $D \ge 2$, with a constant $c > 0$ depending only on D. The detailed proof will appear elsewhere.

Since $v = \frac{D_f}{\delta} = 2\frac{\log N}{\log(N+2)}$, that is, $v = 2\frac{\log 3}{\log 5}$ for $D = 2$, we then get the characterization of the spectral dimension mentioned in the introduction:

COROLLARY *The spectral dimension of K is the homogeneous dimension of K, $D_s = v$.*

5. Poincaré inequalities

Our next result is to show that the homogeneous structure of K is related to the Dirichlet energy form E by a family of scaled Poincaré inequalities on the quasi-balls of K:

THEOREM 2 *Let $D = 2$. The following inequalities are satisfied:*

$$\int_{B(x,r)} \left| u - \overline{u}_{B(x,r)} \right|^2 \mu(\,\mathrm{d}x) \le c_1 r^2 \int_{B(x,qr)} \alpha(u,u)(\,\mathrm{d}x)$$

for every $u \in \mathcal{F}$ and for every $0 < r \le 1$, where $c_1 > 0$ and $q \ge 1$ are suitable constants.

By α we denote above the local Le Jan's energy measure of the Dirichlet form E.

The proof of this theorem, in addition to the tools used in the proof of Theorem 1, exploits the scaling properties of the form E and the imbedding of the domain $\mathcal{F}$ of E into $C(K)$, as given in [FS] and [Ku2]. The detailed proof will appear elsewhere. We expect this theorem too to hold in all dimensions $D \geq 2$, with constants c_1 and q depending on D.

6. The Morrey imbedding

The properties established by Theorem 1 and Theorem 2 allow us to apply the theory of Morrey-Sobolev imbeddings for Dirichlet forms on homogeneous spaces, recently developed in [BM1], [BM2].

In particular, since the homogeneous dimension of K is now $v < 2$, we obtain from [BM1], that the domain $\mathcal{F}$ of E is imbedded in the space of Hölder continuous functions on K of exponent $\beta = 1 - \left(\frac{v}{2}\right)$ with respect to the intrinsic distance d. More precisely, we find that for every $x \in K$ and every $0 < R \leq 1$, the following inequalities hold:

$$|u(y) - u(z)| \leq c \left[\frac{R^2}{\mu(B(x,R))} \int_{B(x,qR)} \alpha(u,u)(\mathrm{d}x) \right]^{1/2} \mathrm{d}(y,z)^{1-v/2} , \qquad \forall y, z \in B(x,R) ,$$

$$|u(y)| \leq c \left[\frac{R^2}{\mu(B(x,R))} \int_{B(x,qR)} \alpha(u,u)(\mathrm{d}x) \right]^{1/2} , \quad \forall y \in B(x,R) ,$$

with suitable constants $c > 0$ and $q \geq 1$ depending only on D.

If $D = 2$, then the exponent δ occurring in the definition of the quasi-distance d is given by $\delta = \frac{\log 5}{\log 4}$ and the dimension v is given by $v = 2\frac{\log 3}{\log 5}$. Therefore, u is Hölder continous with respect to the Euclidean distance, with exponent $\beta_\varepsilon = \delta\beta = \delta\left(1 - \left(\frac{v}{2}\right)\right) = \frac{\log\left(\frac{5}{3}\right)}{\log 4}$, in agreement with Kozlov's result mentioned in the introduction.

Certain asymptotic spectral properties (the so-called *Lenz shift* of the spectrum), related to the action of a homogeneous group (the Heisenberg group), are studied in the paper [BMT]. Again, the homogeneous dimension plays a fundamental role in explaining critical sizes and relevant constants occurring in that problem. This is a further hint that homogeneous structures can be a useful tool in dealing with degenerate or fractal operators.

References

[BP] Barlow M.T. and Perkins E.A., *Brownian motion on the Sierpinski gasket,* Prob. Theo. Rel. Fields **79** (1988), 543–624.

[BM1] Biroli M. and Mosco U., *Sobolev and isoperimetric inequalities for Dirichlet forms on homogeneous spaces,* Atti Accad. Lincei Rend., 1995, to appear.

[BM2] Biroli M. and Mosco U., *Sobolev inequalities on homogeneous spaces,* Sem. Mat. Modena, 1995, to appear.

[BMT] Biroli M., Mosco U. and Tchou N., *Homogenization by the Heisenberg group*, to appear.

[CW] Coifman R.R. and Weiss G., *Analyse harmonique non-commutative sur certains espaces homogènes,* Lectures Notes in Math. **242**, Springer V., 1971.

[F] Fukushima M., *Dirichlet forms, diffusion processes and spectral dimension for nested fractals,* in "Ideas and Methods in Math. Anal., Stochastics and Appl.", S. Albeverio et al. eds., Cambridge Univ. Press, 1992, 151–161.

[FS] Fukushima M. and Shima T., *On a spectral analysis for the Sierpinski gasket,* Potential Analysis **1** (1992), 1–35.

[H] Hutchinson J.E., *Fractals and self similarity,* Indiana Univ. Math. J. **30** (1981), 713–747.

[Ki] Kigami J., *A harmonic calculus on the Sierpinski spaces,* Japan J. Appl. Math. **6** (1989), 259–290.

[Ko] Kozlov S.M., *Harmonization and homogenization on fractals,* Commun. Math. Phys. **153** (1993), 339–357.

[Ku1] Kusuoka S., *A diffusion process on a fractal,* in "Probability Methods in Mathematical Physics", Proc. Taniguchi Int. Symp. (Katata and Kyoto), eds. K. Ito and N. Ikeda, Kinookuniya and North Holland, 1987, pp. 251–274.

[K2] Kusuoka S., *Diffusion processes in nested fractals,* Lect. Notes in Math. 1567, Springer V., 1993.

[L] Lindstrøm T., *Brownian motion on nested fractals,* Memoirs AMS, n. 420, **83** (1990).

[M] Mosco U., *Composite media and asymptotic Dirichlet forms,* J. Funct. Anal. **123** (1994), 368–421.

[S] Stein E.M., *Harmonic analysis,* Princeton Univ. Series, 1994.

Progress in Nonlinear Differential Equations
and Their Applications, Vol. 25

Variational Techniques for Problems in Materials Science

IRENE FONSECA

Department of Mathematics, Carnegie Mellon University
Pittsburgh, PA 15213

ABSTRACT. In recent years there has been a remarkable progress in the mathematical understanding of variational principles for unstable materials phenomena. In this paper some of the techniques developed are outlined.

1. Introduction

In recent years there has been a remarkable progress in the understanding of nonconvex variational problems, motivated in part by ongoing research in materials science. In this paper we will indicate briefly some of the techniques developed and we will discuss the results obtained.

Among the many underlying physically relevant frameworks, we are particularly interested in the applications of nonconvex variational principles to the study of phase transitions, crystals with defects (such as dislocations), metastable equilibrium states for crystals, the onset of microstructure and the creation of concentrations, interactions between fracture and damage, image segmentation, and the behavior and defects of liquid crystals (see [3, 7, 10, 14, 15, 26, 39, 40, 41, 42, 58, 59]). Also, amidst the very many powerful mathematical tools recently developed, here we will concentrate on the study of lower semicontinuity and relaxation techniques (see [11, 12, 25, 29, 33, 55, 56, 57]). Part of the analysis will be undertaken in situations where the growth and coerciveness properties of the energy density are not compatible with the use of standard relaxation techniques (see Section 2; see also [1, 2, 27, 28, 30, 34, 43, 45, 48, 49, 50, 52, 53, 54]). As it turns out, this is the natural setting for the study of some coherent phase transformations and cavitation phenomena. Also, in order to accomodate the study of incoherent phase transformations, image processing, etc, we consider functional spaces that allow for discontinuities of the admissible fields (see [4, 5, 6, 13, 17, 18, 19, 20, 22, 23, 24, 46]).

The study of phase transformations for nonlinear elastic materials suggestes the extension of the analysis to energy functionals involving higher order derivatives, interfacial energies, and discontinuous (Carathéodory) energy densities. The role played by interfacial energy terms in stabilizing the oscillations developed, as well as the dynamical creation of microstructure, concentrations, their evolution and interaction, is very complex and very little is known on this direction. Surface energy terms may be produced naturally by the bulk energy (see [6, 46]), while in some other models interfacial energies are present in the model from the start (see [3, 13, 39]). In the latter case, one may ask what will be the interaction, if any, between the surface energy provided in the model and the one which underlies the bulk energy term (see Section 3).

The framework of generalized measure-valued solutions may turn out to give some insight into these questions. Thermochemical equilibria for coherent two-phase alloys have been analyzed using Young measures (see [42]), and the underlying framework requires a good understanding of constrained variational problems. This remains a virtually unexplored area of research (see [31, 44]).

A common feature of the above described problems is the study of equilibria through a minimization problem of the type

$$\underset{u \in \mathcal{A}}{\text{minimize}}\, I(u),$$

where

$$I(u) := \int_\Omega f(x, u, \nabla u)\, dx + \dots\,, \tag{1.1}$$

and where the nonspecified terms in the energy $I(\cdot)$ account for body forces, lower order terms, interfacial energies, surface energies, etc. Here $\Omega \subset \mathbf{R}^N$ is an open, bounded domain, it represents the reference configuration of a certain material body, and $u : \Omega \to \mathbf{R}^d$ stands for the deformation, mass density, etc.

The main question of the calculus of variations focus on the search of necessary and sufficient conditions satisfied by f which will guarantee (weak) lower semicontinuity of the energy $I(\cdot)$ in some functional space. Usually, due to coercivity and growth conditions verified by f, this space turns out to be a Sobolev space $W^{1,p}$. It is well known (see [2, 11, 55, 57]) that if

$$0 \leq f(x, u, \xi) \leq C(1 + |x|^q) \tag{1.2}$$

and if $p \geq q$, then

$$E(u) := \int_\Omega f(x, u, \nabla u)\, dx \tag{1.3}$$

is $W^{1,p}$-weak lower semicontinuous,

$$u_n \rightharpoonup u \text{ in } W^{1,p} \Rightarrow E(u) \leq \liminf_{n \to \infty} E(u_n),$$

if and only if $f(x, u, \cdot)$ is *quasiconvex*, i.e.

$$f(x, u, A) \leq \int_{(0,1)^N} f(x, u, A + \nabla \varphi(y))\, dy$$

for all matrices $A \in \mathbf{M}^{d \times N}$ and all $\varphi \in W_0^{1,p}((0,1)^N, \mathbf{R}^d)$. Here, and in what follows, $\mathbf{M}^{d \times N}$ denotes the vector space of $d \times N$ real-valued matrices. We recall that in the scalar case ($N = 1$ or $d = 1$) a function is quasiconvex if and only if is convex. In cases where f is not quasiconvex, and in order to study limiting energies and stress states of the system, we consider the *relaxed energy*

$$\mathcal{F}(u, \Omega) := \inf_{\{u_n\}} \left\{ \liminf_{n \to \infty} \int_\Omega f(x, u_n, \nabla u_n)\, dx : u_n \rightharpoonup u \text{ in } W^{1,p} \right\}.$$

One of the central questions of the relaxation theory is to find the new relaxed bulk energy density $\overline{f}$, i.e. an integral representation for $\mathcal{F}(u,\cdot)$ of the form

$$\mathcal{F}(u,\Omega) := \int_\Omega \overline{f}(x,u,\nabla u)\,dx.$$

We recall that, under suitable regularity and growth assumptions, $\mathcal{F}(\cdot,\Omega)$ is now a lower semicontinuous functional, and the relaxation theory (see [29]) guarantees that

$$\inf_{u\in\mathcal{A}} \int_\Omega f(x,u,\nabla u)\,dx = \min_{u\in\mathcal{A}} \int_\Omega \overline{f}(x,u,\nabla u)\,dx,$$

where $\overline{f}(x,u,\cdot)$ is the *quasiconvexification* $Qf(x,u,\cdot)$ of $f(x,u,\cdot)$, precisely

$$Qf(x,u,A) := \inf\left\{\int_{(0,1)^N} f(x,u,A+\nabla\varphi(y))\,dy : \varphi \in W_0^{1,p}\left((0,1)^N,\mathbf{R}^d\right)\right\}.$$

Often, the regularity or the growth conditions required to apply the above standard relaxation techniques fail. In this paper we will overview new mathematical approaches which will allow us to study some of the situations not covered under the usual assumptions. In Section 2 we concentrate on the analysis of the bulk energy term, precisely on the lower semicontinuity properties of (1.3) when the function f satisfies the bound (1.2), and the relaxation in taken among sequences $u_n \in W^{1,q}(\Omega,\mathbf{R}^d)$, $u_n \to u$ in $W^{1,p}$ weak, with $p<q$. On Section 3 we extend the analysis to include the relaxation in BV of the same functionals in the case where coerciveness only guarantees $W^{1,1}$ bounds for minimizing sequences. Section 4 addresses the relaxation in BV of energies of the type (1.1) which involve bulk and interfacial energy contributions, and where the energy densities have linear growth. The analysis of the superlinear growth case is undertaken partially on Section 5.

2. Growth conditions versus Sobolev bounds

In collaboration with Marcellini [45] and with Malý [43], we studied $W^{1,p}$-weak lower semicontinuity properties of the functional $E(\cdot)$ introduced in (1.3), where f satisfies (1.2) and $p<q$. A prototype example is given by

$$G(u) := \int_\Omega \left(h(\nabla u) + g(\det\nabla u)\right)\,dx\,, \tag{2.1}$$

where

$$\frac{1}{C_1}\,|\xi|^p - C_2 \le h(\xi) \le C_1(1+|\xi|^p)\,,\ \frac{1}{C_1}\,|t| - C_2 \le g(t) \le C_1(1+|t|)$$

for some $C_1>0, C_2\ge 0, N-1<p<N$ and for all $\xi\in\mathbf{M}^{N\times N}, t\in\mathbf{R}$. Integrands of this type are considered in nonlinear elasticity and the condition $p<q=N$ plays a fundamental role in cavitation analysis (see [11]).

The (weak) lower semicontinuity problem in $W^{1,p}$ for *polyconvex* integrands (i.e., suprema of rank-one affine functions) and for p below the growth exponent q

was first considered by Marcellini [55]. In particular, if we restrict to our prototype example (2.1), Marcellini proved the lower semicontinuity for $p > \frac{N^2}{N+1}$. This result was extended to the case $p > N-1$ by Dacorogna and Marcellini [33]; the borderline case $p = N-1$ was considered in [52] with a partial success, and it completely established by Acerbi, Dal Maso and Sbordone [1, 34] (see also [28, 48, 49]).

With Marcellini [45] we studied a class of quasiconvex integrands $f = f(\xi)$ satisfying some structure conditions naturally verified by (2.1). We proved lowersemicontinuity of the energy (1.3) for $p > q-1$. For non quasiconvex integrands, we define the relaxed energy

$$\mathcal{F}^{q,p}(u,\Omega) := \inf_{\{u_n\}} \left\{ \liminf_{n\to\infty} \int_\Omega f(\nabla u_n) : u_n \in W^{1,q}(\Omega, \mathbf{R}^d), \right.$$
$$\left. u_n \rightharpoonup u \text{ weakly in } W^{1,p}(\Omega,\mathbf{R}^d) \right\}$$

and we proved that

$$\mathcal{F}^{q,p}(u) \geq \int_\Omega QW(\nabla u(x))\, dx.$$

Equality holds provided $u \in W^{1,q}(\Omega;\mathbf{R}^d)$. Recently, Malý [54] extended the later result to the borderline case $p = q-1$, assuming always suplementary structural conditions on f.

In collaboration with Malý [43] we developed a new variational approach which allows us to eliminate the above mentioned additional structural assumptions if the growth condition is (1.2) and $p > q\frac{N-1}{N}$. We defined a new relaxed energy

$$\mathcal{F}^{q,p}_{\text{loc}}(u,\Omega) := \inf_{\{u_n\}} \left\{ \liminf_{n\to\infty} \int_\Omega f(x,u_n,\nabla u_n) : u_n \in W^{1,q}_{\text{loc}}(\Omega, \mathbf{R}^d), \right.$$
$$\left. u_n \rightharpoonup u \text{ weakly in } W^{1,p}(\Omega,\mathbf{R}^d) \right\}.$$

The behaviors of $\mathcal{F}^{q,p}_{\text{loc}}$ and $\mathcal{F}^{q,p}$ may be rather different, and their values depend, in a strikingly complicated way, on the ranges of p, q, and on the regularity properties of u. Consider the example where $N = d$, $f(x,u,\xi) = f(\xi) = |\det \xi|$. Notice that f is polyconvex and the growth condition

$$0 \leq f(\xi) \leq |\xi|^N$$

is satisfied. It is well known that

$$\mathcal{F}^{q,p}(u,\Omega) \geq \int_\Omega |\det \nabla u|\, dx \tag{2.2}$$

if $p, q \geq N$. Recently, this result has been improved to include the case where $q \geq N$ and $p \geq N-1$ (see [28]). If $u \in W^{1,N}(\Omega,\mathbf{R}^N)$, then we get equality in (2.2), whereas for $u \notin W^{1,N}(\Omega,\mathbf{R}^N)$ it is difficult to describe $\mathcal{F}^{q,p}(u,\Omega)$. We obtain

$$\mathcal{F}^{q,p}(u,\Omega) = 0 \tag{2.3}$$

if $q < N$ (see [12]) or if $p < N-1$ (see [43, 52]).

As before, in order to identify $\mathcal{F}^{q,p}(u,\Omega)$ and $\mathcal{F}^{q,p}_{\mathrm{loc}}(u,\Omega)$, in [43] we started by obtaining a lower bound for the relaxed energy,

$$\mathcal{F}^{q,p}(u,\Omega) \geq \int_\Omega Qf(x,u,\nabla u)\,dx, \tag{2.4}$$

which amounts to proving a lower semicontinuity result for quasiconvex integrals, namely

$$\liminf_{n\to\infty} \int_\Omega f(x,u_n,\nabla u_n)\,dx \geq \int_\Omega f(x,u,\nabla u)\,dx \tag{2.5}$$

if $u \in W^{1,p}(\Omega,\mathbf{R}^d)$, $u_n \in W^{1,q}(\Omega,\mathbf{R}^d)$, $u_n \rightharpoonup u$ weakly in $W^{1,p}(\Omega,\mathbf{R}^d)$, $f(x,u,\cdot)$ is quasiconvex, and $p > q\frac{N-1}{N}$. In view of (2.3), the inequality (2.5) may no longer be valid if $p < q\frac{N-1}{N}$.

Further, we investigated the dependence of $\mathcal{F}^{q,p}(u,U)$ and $\mathcal{F}^{q,p}_{\mathrm{loc}}(u,U)$ on the open subsets $U \subset \Omega$. Assuming that

$$0 \leq f(x,\zeta,\xi) \leq C(1+|\zeta|^r+|\xi|^q),$$

we proved that if $p > \max\left\{q\frac{N-1}{N}, \frac{rN^2-rN}{N^2+rN-r}\right\}$, and if $\mathcal{F}^{q,p}(u,\Omega) < \infty$, then there exists a finite, nonnegative, Radon measure μ such that

$$\mathcal{F}^{q,p}(u,U) = \mu(U) \tag{2.6}$$

for all open sets $U \subset \Omega$ with $\mu(\partial U) = 0$. In addition,

$$\mathcal{F}^{q,p}_{\mathrm{loc}}(u,U) = \int_U Qf(x,u,\nabla u)\,dx + \lambda(U)$$

holds for all open sets $U \subset \Omega$, where λ is some finite, nonnegative, Radon measure.

The representation formula (2.6) may fail if $p \leq q\frac{N-1}{N}$, as illustrated by an example of the class (2.1), provided by Acerbi and Dal Maso [1]: if $f(\xi) := |\xi|^{N-1} + |\det \xi|$ and setting $p = N-1$, then $\mathcal{F}^{q,p}(u,\cdot)$ is not even subadditive. Here $q = d = N$, $r = p = N-1$, $u(x) = \frac{x}{|x|} \in W^{1,s}(B,\mathbf{R}^N)$ for all $s < N$, in particular for $s = p$. Then

$$\rho \mapsto \mathcal{F}^{q,p}(u,\rho B) - \int_{\rho B} f(\nabla u)$$

is of order ρ at 0, whereas

$$\mathcal{F}^{q,p}(u,B\setminus\rho B) - \int_{B\setminus\rho B} f(\nabla u) = 0.$$

Hence $\mathcal{F}^{q,p}(u,\cdot)$ is not be subadditive.

In this example, the additivity property failed due to the fact that $p \leq q\frac{N-1}{N}$. Now we will see that, in spite of requiring $p > q\frac{N-1}{N}$, the measure representation (2.6) may not hold for open sets U with $\mu(\partial U) > 0$. Let $q = d = N$ and $u(x) := \frac{x}{|x|}$, but now $N > p > N-1$, and

$$f(\xi) := |\det \xi|\,.$$

Let $\mu := \mathcal{L}^N(B)\,\delta_0$ be the $\mathcal{L}^N(B)$-multiple of the Dirac measure at 0. Then (see [45], Theorem 4.1),

$$\mathcal{F}^{q,p}(u,U) = \mu(U) \tag{2.7}$$

if $\mu(\partial U) = 0$. For $U = \{x \in B : x_1 > 0\}$ we have

$$\mathcal{F}^{q,p}(u,U) = \mu(U) < \mu(\overline{U}), \tag{2.8}$$

and if $U := B \setminus \{0\}$ then

$$\mathcal{F}^{q,p}(u,U) = \mu(B) = \mu(\overline{U}) > \mu(U)\,,$$

because each $v \in W^{1,q}(U)$ is also in $W^{1,q}(B)$ (the point 0 is a removable singularity). Clearly, $\mathcal{F}^{q,p}(u,\cdot)$ cannot be a measure since in this case, and by (2.7), it would have to be the measure μ, contradicting (2.8).

Moreover, we showed that (2.6) holds for all open sets $U \subset \Omega$ if and only if it exists a Radon measure ν such that

$$\mathcal{F}^{q,p}(u,U) \le \nu(U)$$

for all open subset $U \subset \Omega$.

Under standard continuity hypotheses on f with respect to x and ζ (see [13]), the lower semicontinuity result (2.4) implies the estimate

$$\mu_a \ge Qf(x,u,\nabla u)\mathcal{L}^N$$

for the absolutely continuous part μ_a of μ, where $\mathcal{L}^N$ is the N-dimensional Lebesgue measure. Actually, in all known examples the equality

$$\mu_a = Qf(x,u,\nabla u)\,\mathcal{L}^N$$

holds.

The main novelty of the paper [43] lies on the construction of a linear operator Tu from $W^{1,p}$ into $W^{1,q}$ which conserves boundary values and improves integrability of u and ∇u. Namely, the $W^{1,q}$-norm of Tu is estimated in terms of a special maximal function if $p > q\frac{N-1}{N}$. We use this extension operator to "connect" two functions across a thin transition layer and to estimate the increase of the energy. We remark that the standard way to perform this connection, by means of convex combinations using cut-off functions, would not achieve a comparable result, namely an arbitrarily small increase of the energy on an arbitrarily thin transition layer, since the admissible sequences may not remain bounded in $W^{1,q}(\Omega,\mathbf{R}^d)$.

3. Interactions between bulk and interfacial energies: bulk generated surface energy

Several problems in Mathematical Physics may be modelled by functionals of the form (1.1), where the underlying function spaces should allow discontinuous vector-valued functions u. In particular, relaxation techniques for these energy functionals have been used in the study of variational models in fracture mechanics, where u is the displacement field and the discontinuity set $S(u)$ represents the crack site (see [26, 39]); in the theory of computer vision and image segmentation, where the function u represents the so-called image reconstruction and $S(u)$ is the edge contour (see [3, 4, 5, 7, 36, 58, 59]). As a common feature to these problems, we seek to minimize (1.1) on the class BV of functions of bounded variation, or on some suitable subclass of it.

We recall that the distributional derivative of a function u in $BV(\Omega;\mathbf{R}^d)$ is a finite Radon measure on Ω, and it may be decomposed as

$$Du = \nabla u\, \mathcal{L}^N + (u^+ - u^-) \otimes \nu\, H^{N-1} \lfloor S(u) + C(u),$$

where ∇u is a $L^1(\Omega;\mathbf{M}^{d\times N})$ function (the density of the absolutely continuous part of the measure Du), $S(u)$, the jump set of u, is an $N-1$ rectifiable hypersurface with normal vector ν, u^+ and u^-, the traces of u on each side of $S(u)$, are such that, for H^{N-1}-a.e. x_0 in $S(u)$,

$$\lim_{\epsilon\to 0^+} \frac{1}{\epsilon^N} \int_{\{y\in B(x_0,\epsilon);(y-x_0)\cdot\nu_{x_0}>(<)0\}} |u(y) - u^+(u^-)(x_0)|^{N/(N-1)} dy = 0, \quad (3.1)$$

and $C(u)$, the Cantor part of the measure Du, satisfies

$$H^{N-1}(B) < +\infty \Rightarrow |C(u)|(B) = 0$$

for any Borel subset B of Ω (see e.g. [37, 38, 60]).

In collaboration with Müller [46] (see also [6]), we studied the lower semicontinuity and relaxation properties of bulk energies, i.e. when (1.1) reduces to (1.3), and in the case where the density $f(x,u,\cdot)$ has linear growth. Clearly, the natural space for performing the relaxation is BV, rather than $W^{1,1}$. This study was undertaken in the scalar case by Dal Maso [32].

Let $u \in BV(\Omega,\mathbf{R}^d)$, and let $\mathcal{F}(u,\cdot)$ be the relaxation in BV of the functional (1.3), i.e.

$$\mathcal{F}(u,U) := \inf_{\{u_n\}} \left\{ \liminf_{n\to\infty} \int_U f(x,u_n,\nabla u_n)\, dx : u_n \rightharpoonup u \text{ in } L^1(\Omega,\mathbf{R}^d) \right\},$$

where U is an open subset of Ω. Under suitable continuity hypotheses on f, and assuming that $c\,g(x,u)(1+||\xi||) \le f(x,u,\xi) \le C\,g(x,u)(1+||\xi||)$ for some non-negative function g, we obtained the following representation

$$\begin{aligned} \mathcal{F}(u) &= \int_\Omega Qf(x,u(x),\nabla u(x))\, dx \\ &+ \int_{S(u)} K_f(x,u^-(x),u^+(x),\nu(x)) dH^{N-1}(x) \\ &+ \int_\Omega (Qf)^\infty(x,u(x),dC(u)). \end{aligned}$$

The surface energy density $K_f : \Omega \times \mathbf{R}^d \times \mathbf{R}^d \times S^{N-1} \to [0,+\infty)$ is given by (see also [21, 24])

$$K_f(x,a,b,\nu) = \inf\left\{\int_{Q_\nu} f^\infty(x,v(y),\nabla v(y))dy : v \in \mathcal{A}(a,b,\nu)\right\} \qquad (3.2)$$

where Q_ν is a unit cube centered at the origin with two of its faces normal to ν, and

$$\mathcal{A}(a,b,\nu) := \Big\{v \in W^{1,1}(Q_\nu;\mathbf{R}^d) \quad : \quad v(y) = a \text{ if } y \in \partial Q_\nu,\ y\cdot\nu \leq 0, \\ v(y) = b \text{ if } y \in \partial Q_\nu,\ y\cdot\nu \geq 0\Big\}.$$

If f does not depend explicitly on u, then it turns out that

$$K_f(x,a,b,\nu) = (Qf)^\infty(x,(b-a)\otimes\nu)$$

(see [6] and [46], Remark 2.17), where $(Qf)^\infty$ is the *recession function* of f, namely

$$(Qf)^\infty(x,A) := \limsup_{n\to\infty}\frac{QF(x,tA)}{t}.$$

The method introduced to obtain this integral representation, the *blow-up method*, may be summarized as follows. Suppose that we want to represent $\mathcal{F}(u,\Omega)$ as

$$\begin{aligned}\mathcal{F}(u,\Omega) &= \int_\Omega \overline{f}(x,u,\nabla u), dx \\ &\quad + \int_{S(u)} K_f(x,u^-(x),u^+(x),\nu(x))dH^{N-1}(x) \\ &\quad + \int_\Omega G(x,u(x),dC(u)). \end{aligned} \qquad (3.3)$$

In order to identify the energy densities $\overline{f}, K_f, G$, we claim that it suffices to characterize $\mathcal{F}(v,Q)$ when Q is a unit cube and v is obtained as the blow-up around a point x_0 of the function u. Precisely, let $\{u_n\}$ be a minimizing sequence for $\mathcal{F}(u,\Omega)$, i.e. $u_n \to u$ in L^1 and

$$\mathcal{F}(u,\Omega) = \lim_{n\to\infty}\int_\Omega f(x,u_n,\nabla u_n)\,dx,$$

and define the sequence $\{\mu_n\}$ of Radon measures by

$$\mu_n := f(x,u_n,\nabla u_n)\mathcal{L}^N.$$

Assuming that $\mathcal{F}(u,\Omega) < +\infty$, it follows that $\sup_n |\mu_n|(\Omega) < +\infty$ and so there exists a subsequence (still denoted μ_n) and a finite Radon measure μ such that $\mu_n \overset{*}{\rightharpoonup} \mu$ in the sense of measures, i.e. for every $\psi \in C_0(\Omega)$

$$\lim_{n\to+\infty}\int_\Omega \psi(x)f(x,u_n(x),\nabla u_n(x))\,dx = \int_\Omega \psi(x)d\mu(x).$$

By the Radon-Nikodym Theorem, we may decompose μ as the sum of four mutually singular nonnegative measures

$$\mu = \mu_a \mathcal{L}^N + \mu_J H^{N-1} \lfloor S(u) + \mu_c |C(u)| + \mu_s.$$

The equality (3.3) is achieved provided one can prove that

$$\begin{aligned}
&\mu_a(x_0) = \overline{f}(x_0, u(x_0), \nabla u(x_0)) \text{ for } \mathcal{L}^N \text{ a.e. } x_0 \in \Omega,\\
&\mu_J(x_0) = K_f(x_0, u^-(x_0), u^+(x_0), \nu(x_0)) \text{ for } H_{N-1} \text{ a.e. } x_0 \in \Omega \cap S(u),\\
&\mu_c(x_0) = G\left(x_0, u(x_0), \frac{dC(u)}{d|C(u)|}(x_0)\right) \text{ for } |C(u)| \text{ a.e. } x_0 \in \Omega,\\
&\mu_s = 0.
\end{aligned}$$

To show that

$$\mu_a(x_0) = \overline{f}(x_0, u(x_0), \nabla u(x_0)) \text{ for } \mathcal{L}^N \text{ a.e. } x_0 \in \Omega,$$

we select a point $x_0 \in \Omega$ such that the following hold:

$$\lim_{\epsilon \to 0^+} \frac{1}{\epsilon} \int_{B(x_0,\epsilon)} |u(x) - u(x_0) - \nabla u(x_0) \cdot (x - x_0)| \, dx = 0, \tag{3.4}$$

$$\mu_a(x_0) = \lim_{\epsilon \to 0^+} \frac{\mu(Q(x_0,\epsilon))}{\mathcal{L}^N(Q(x_0,\epsilon))} \text{ exists and is finite }, \tag{3.5}$$

where $Q(x_0, \epsilon) := x_0 + \epsilon Q, Q = (0,1)^N$. Denote by D the (at most countable) set of all $\epsilon > 0$ such that the boundary of $x_0 + \epsilon Q$ is not μ-negligible. For every $\epsilon \notin D$, we have

$$\begin{aligned}
\mu_a(x_0) &= \lim_{\epsilon \to 0^+, \, \epsilon \notin D} \frac{\mu(Q(x_0,\epsilon))}{\epsilon^N}\\
&= \lim_{\epsilon} \frac{1}{\epsilon^N} \lim_{n \to +\infty} \int_{x_0 + \epsilon Q} f(x, u_n(x), \nabla u_n(x)) \, dx\\
&= \lim_{\epsilon} \lim_{n \to +\infty} \int_Q f(x_0 + \epsilon y, u_n(x_0 + \epsilon y), \nabla u_n(x_0 + \epsilon y)) dy\\
&= \lim_{\epsilon} \lim_{n \to +\infty} \int_Q f(x_0 + \epsilon y, u_{n,\epsilon}(y), \nabla u_{n,\epsilon}(y)) dy,
\end{aligned}$$

where

$$u_{n,\epsilon}(y) := \frac{u_n(x_0 + \epsilon y) - u(x_0)}{\epsilon}.$$

Clearly, by (3.4)

$$\lim_{\epsilon \to 0^+} \lim_{n \to +\infty} ||u_{n,\epsilon} - u_0||_{L^1(Q)} = 0$$

with $u_0(y) := \nabla u(x_0) y$, and so, due to the continuity of f, and after extracting a diagonal subsequence, we conclude that

$$\mu_a(x_0) = \lim_{k \to \infty} \int_Q f(x_0, u(x_0), \nabla v_k(y)) \, dy$$

where v_k is a sequence bounded in $W^{1,1}(Q,\mathbf{R}^d)$ such that $v_k \to u_0$ in L^1. The advantage of this method is that we reduced the relaxation problem for an arbitrary BV function to the relaxation problem when the target is a homogeneous function. In the same way, when searching for a characterization of the interfacial energy density K_f, the blow-up method, together with (3.1), allows us to focus on the case where the macroscopic limit is the piecewise constant function

$$u_1(y) := \begin{cases} u^+(x_0) & \text{if } y.\nu(x_0) > 0 \\ u^-(x_0) & \text{if } y.\nu(x_0) < 0. \end{cases}$$

The density G on the Cantor part is obtained following a similar argument.

4. Interactions between bulk and interfacial energies: the linear growth case

The next step was taken in collaboration with Barroso, Bouchitté and Buttazzo [13], where we studied the relaxation of (1.1) in the case where there is an interfacial contribution from the start. Precisely, we studied the relaxation $\mathcal{F}(u,\Omega)$ with respect to the L^1 convergence of the functional defined in $SBV(\Omega;\mathbf{R}^d)$ by

$$u \mapsto I(u) := \int_\Omega f(x,\nabla u(x))\,dx + \int_{S(u)} \varphi(x,[u](x),\nu(x))dH^{N-1}(x) \tag{4.1}$$

where we assume that $f(x,\cdot)$ is quasiconvex and has linear growth, and that $\varphi(x,\cdot,\nu)$ grows also linearly,

$$c||A|| \leq f(x,A) \leq C(1+||A||), \quad c_1|\xi| \leq \varphi(x,\xi,\nu) \leq C_1|\xi|. \tag{4.2}$$

The space $SBV(\Omega;\mathbf{R}^d)$ of *special functions of bounded variation* was firstly introduced in [35]; a function $u \in BV(\Omega;\mathbf{R}^d)$ is said to be of special bounded variation if $C(u) = 0$, i.e. the distributional derivative of u can be written as $Du = \nabla u\,\mathcal{L}^N + (u^+ - u^-)\otimes \nu H^{N-1}\lfloor S(u)$. Under some technical continuity conditions and using the blow-up method described in Section 3, we obtained the integral representation

$$\begin{aligned}\mathcal{F}(u,\Omega) &= \int_\Omega g(x,\nabla u(x))\,dx \\ &\quad + \int_{S(u)} h(x,[u](x),\nu(x))dH^{N-1}(x) \\ &\quad + \int_\Omega g^\infty(x,dC(u))\end{aligned}$$

in $BV(\Omega;\mathbf{R}^d)$, where g is the quasiconvexification of the *inf-convolution* of f and φ_0, $g = Q(f\nabla\varphi_0)$, the inf-convolution is defined by

$$f\nabla\varphi_0(x,A) := \inf\left\{f(x,A-a\otimes b)+\varphi_0(x,a,b) : a\in\mathbf{R}^d, b\in S^{N-1}\right\},$$

and h is given by

$$h(x_0, \eta, \nu) := \inf \left\{ \int_{Q_\nu} f^\infty(x_0, \nabla u(x))\, dx + \int_{S(u)\cap Q_\nu} \varphi(x_0, [u](x), \nu(x)) dH^{N-1}(x) : u \in \mathcal{A}(\eta, \nu) \right\}.$$

Here Q_ν is any unit cube centered at the origin and with two of its faces normal to ν,

$$\begin{aligned} \mathcal{A}(\eta, \nu) \quad := \quad & \Big\{ v \in SBV_{loc}(S_\nu; \mathbf{R}^d) : v(y) = 0 \text{ if } y \cdot \nu = -\frac{1}{2}, \\ & v(y) = \eta \text{ if } y \cdot \nu = \frac{1}{2}, \\ & v \text{ is 1-periodic in the directions of } \nu_1, \dots, \nu_{N-1} \Big\}, \end{aligned}$$

$\{\nu_1, \dots, \nu_{N-1}, \nu\}$ forms an orthonormal basis of $\mathbf{R}^N$, and S_ν stands for the strip $S_\nu := \left\{ y \in \mathbf{R}^N : |y \cdot \nu| < \frac{1}{2} \right\}$. In the above, as usual f^∞ (resp. g^∞) denotes the recession function of f (resp. g), and φ_0 is the positively homogeneous of degree one function defined by

$$\varphi_0(x, \eta, \nu) := \lim_{t \to 0^+} \frac{\varphi(x, t\eta, \nu)}{t}.$$

Bouchitté, Braides and Buttazzo [16] studied the scalar version of this problem, and they extended the results to the case of linear growth in f or φ, although they needed to make an isotropy assumption. Recently, Braides and Coscia [22] obtained an integral representation of the relaxation with respect to the L^1 topology of the functional

$$u \mapsto \int_\Omega f(\nabla u(x))\, dx + \int_{S(u)} \varphi([u](x) \otimes \nu(x)) dH_{N-1}(x),$$

under the assumption that φ is positively homogeneous of degree one. Very mild restrictions are placed on f.

5. Interactions between bulk and interfacial energies: the superlinear growth case

The analysis in [13] was carried out under the linear growth assumptions (4.2). These prove to be too exclusive in some settings. As an example, in problems in image segmentation and interaction between fracture and damage, it is commonly assumed that the bulk energy density $f(x, u, \cdot)$ grows superlinearly.

In the study of variational models in fracture mechanics, u is the displacement field and $S(u)$ represents the fracture region (see [26, 39]). In collaboration with Francfort [39] we studied variationally equilibria for materials that experienced brutal partial damage. To this end, we used the model introduced in [47], where it is assumed that the material is only allowed to brutally drop from its healthy state to its damaged state, the latter retaining some positive definite stiffness. Further, the material may undergo fracture, in which case material discontinuities at a macroscopic level will develop. The quasistatic evolution of both damage and fracture is governed by a yield criterion, commonly accepted in fracture mechanics. After the work of Griffith (see [51], Chapter 4), the criterion compares the decrease in potential energy due to either damage (in which case it is a local decrease) or fracture (in which case it is a global decrease) to the resulting increment of energy dissipated through either process. Both processes are further assumed to be irreversible. In other words, self-healing is absent from both the damaged part of the material and the cracks through that material.

The adopted model results in a time indexed sequence of partial minimization problems. At each time step the potential energy to be minimized is of the form

$$E(u,\Omega) := \int_\Omega W(\nabla u)\,dx + \lambda H^{N-1}(S(u)) - \int_\Omega f\cdot u\,dx,$$

where $W(\xi)$ is the "elastic" energy, $\lambda > 0$ is a dissipation rate and f represents the body loadings. The density W is non quasiconvex, and it satisfies

$$\alpha|\xi|^p \le W(\xi) \le \beta(\delta + |\xi|^p),\ \xi \in \mathbf{M}^{d\times N},$$

$$|W(\xi) - W(\eta)| \le \gamma(1 + |\eta|^{p-1} + |\xi|^{p-1})|\xi - \eta|,\ \xi,\eta \in \mathbf{M}^{d\times N},$$

where $\alpha,\beta,\gamma > 0$, $\delta \ge 0$ and $1 < p < +\infty$. If $u \in SBV(\Omega,\mathbf{R}^d)$, and setting

$$\mathcal{F}(u,U) := \inf_{\{u_n\}} \left\{ \liminf_{n\to\infty} E(u_n,U) : u_n \to u \text{ in } L^1(\Omega,\mathbf{R}^d)\right\},$$

where U is an open subset of Ω, then we showed in [39] that

$$\mathcal{F}(u,U) := \int_U QW(\nabla u)\,dx + H^{N-1}(S(u)\cap U). \tag{5.1}$$

The analysis relies heavily on the blow-up method (see Section 3; see also [46]) and on Ambrosio's lower semi-continuity result in $SBV(\Omega;\mathbf{R}^d)$ for quasiconvex Carathéodory integrands with superlinear growth (see [3]). In [39] we used (5.1) to investigate the quasistatic evolution of damage and fracture at discretized times.

An important question that remains to be answered regards the regularity of the interface $S(u)$ and of u, whenever u minimizes the relaxed energy, and under suitable boundary conditions or constraints. With the exception of the case where $QW(\xi) = |\xi|^p$ (see [8, 9, 36]), this question remained virtually untouched until very recently. Ongoing work on this direction is being carried out in collaboration with Francfort.

Acknowledgements. This work was supported by the Army Research office and the National Science Foundation through the Center for Nonlinear Analysis at Carnegie Mellon University. Also, this research was partially supported by the National Sience Fondation under Grant no. DMS-9201215.

References

[1] Acerbi, E. and G. Dal Maso. New lower semicontinuity results for polyconvex integrals case. Preprint SISSA (Trieste) **52/M** (1993).

[2] Acerbi, E. and N. Fusco. Semicontinuity problems in the Calculus of variations. *Arch. Rat. Mech. Anal.* **86** (1984), 125–145.

[3] Ambrosio, L. On the lower semicontinuity of quasiconvex integrals in $SBV(\Omega; \mathbf{R}^k)$. *Nonlinear Anal.* **23** (1994), 405–425.

[4] Ambrosio, L. and A. Braides. Functionals defined on partitions of sets of finite perimeter, I: Integral representation and Γ-convergence. *J. Math. Pures Appl.* **69** (1990), 285–305.

[5] Ambrosio, L. and A. Braides. Functionals defined on partitions of sets of finite perimeter, II: Semicontinuity, relaxation and homogenization. *J. Math. Pures Appl.* **69** (1990), 307–333.

[6] Ambrosio, L. and G. Dal Maso. On the relaxation in $BV(\Omega; \mathbf{R}^m)$ of quasiconvex integrals. *J. Funct. Anal.* **109** (1992), 76–97.

[7] Ambrosio, L. and E. De Giorgi. Un nuovo tipo di funzionale del calcolo delle variazioni. *Atti. Accad. Naz. Lincei Rend. Cl. Sci. Fis. Mat. Natur.* **82** (1988), 199–210.

[8] Ambrosio, L., N. Fusco and D. Pallara. Partial regularity of free discontinuity sets II. Preprint, U. Pisa, 1994.

[9] Ambrosio, L. and D. Pallara. Partial regularity of free discontinuity sets I. Preprint, U. Pisa, 1994.

[10] Ambrosio, L. and E. Virga. A boundary value problem for nematic liquid crystals with variable degree of orientation. *Arch. Rational Mech. Anal.* **114** (1991), 335–347.

[11] Ball, J. M. Convexity conditions and existence theorems in nonlinear elasticity. *Arch. Rat. Mech. Anal.* **63** (1977), 337–403.

[12] Ball, J. M. and F. Murat, F. $W^{1,p}$ quasiconvexity and variational problems for multiple integrals. *J. Funct. Anal.* **58** (1984), 225–253.

[13] Barroso, A. C., G. Bouchitté, G. Buttazzo and I. Fonseca. Relaxation in $BV(\Omega, \mathbf{R}^p)$ of energies involving bulk and surface energy contributions. *Arch. Rat. Mech. Anal.*, to appear.

[14] Barroso, A. C. and I. Fonseca. Anisotropic singular perturbations. *Proc. Royal Soc. Edin.* **124A** (1994), 527–571.

[15] Blake, A. and A. Zisserman. *Visual Reconstruction*, MIT Press, Boston, 1987.

[16] Bouchitté, G., A. Braides and G. Buttazzo. Relaxation results for some free discontinuity problems. *J. Reine Angew. Math.*, to appear.

[17] Bouchitté, G. and G. Buttazzo. New lower semicontinuity results for non convex functionals defined on measures. *Nonlinear Anal.* **15** (1990), 679–692.

[18] Bouchitté, G. and G. Buttazzo. Integral representation of nonconvex functionals defined on measures. *Ann. Inst. H. Poincaré Anal. Non Linéaire*, **9** (1992), 101–117.

[19] Bouchitté, G. and G. Buttazzo. Relaxation for a class of nonconvex functionals defined on measures. *Ann. Inst. H. Poincaré Anal. Non Linéaire*, **10** (1993), 345–361.

[20] Bouchitté, G. and G. Dal Maso. Integral representation and relaxation of convex local functionals on $BV(\Omega)$. *Ann. Scuola Norm. Sup. Pisa*, **20** (1993), 483–533.

[21] Bouchitté, G., I. Fonseca and L. Mascarenhas. On the relaxation of multiple integrals with Carathéodory integrands. In preparation.

[22] Braides, A. and A. Coscia. The interaction between bulk energy and surface energy in multiple integrals. *Proc. Royal Soc. Edin.*, to appear.

[23] Braides, A. and V. De Cicco. New lower semicontinuity results for functionals defined on BV. Preprint SISSA (Trieste) **66** (1993).

[24] Braides, A. and C. Piat. A derivation formula for integral functionals defined on $BV(\Omega)$. Preprint SISSA (Trieste) **144** (1993).

[25] Buttazzo, G. *Semicontinuity, Relaxation and Integral Representation in the Calculus of Variations.* Pitman Res. Notes Math. Ser. **207**, Longman, Harlow, 1989.

[26] Buttazzo, G. Energies on BV and variational models in fracture mechanics. Preprint Dip. Mat. Univ. Pisa (1994).

[27] Carbone, L. and R. De Arcangelis. Further results on Γ-convergence and lower semicontinuity of integral functionals depending on vector-valued functions. *Richerche Mat.* **39** (1990), 99–129.

[28] Celada, P. and G. Dal Maso. Further remarks on the lower semicontinuity of polyconvex integrals. *Ann. Inst. H. Poincaré Anal. Non Linéaire*, to appear.

[29] Dacorogna, B. *Direct Methods in the Calculus of Variations.* Appl. Math. Sciences **78**, Springer-Verlag, Berlin, 1989.

[30] Dacorogna, B. and P. Marcellini. Semicontinuité pour des intégrandes polyconvexes sans continuité des determinants. *C. R. Acad. Sci. Paris Sér. I Math.* **311,6** (1990), 393–396.

[31] Dacorogna, B., I. Fonseca and V. Mizel. Relaxation problems under constraints. To appear.

[32] Dal Maso, G. Integral representation on $BV(\Omega)$ of Γ-limits of variational integrals. *Manuscripta Math.* **30** (1980), 387–416.

[33] Dal Maso, G. *An Introduction to* Γ*-Convergence.* Birkhäuser, Boston, 1993.

[34] Dal Maso, G. and C. Sbordone. Weak lower semicontinuity of polyconvex integrals: a borderline case. Preprint SISSA (Trieste) **45/M** (1993).

[35] De Giorgi, E. and L. Ambrosio. Un nuovo tipo di funzionale del calcolo delle variazioni. *Atti. Accad. Naz. Lincei Rend. Cl. Sci. Fis. Mat. Natur.* **82** (1988), 199–210.

[36] De Giorgi, E., M. E. Carriero and A. Leacci. Existence theorem for a minimum problem with free discontinuity set. *Arch. Rat, Mech. Anal.* **108** (1989), 195–218.

[37] Evans, L. C. and R. F. Gariepy. *Measure Theory and Fine Properties of Functions.* CRC Press, 1992.

[38] Federer, H. *Geometric Measure Theory.* Springer, 1969.

[39] Fonseca, I. and G. Francfort. Relaxation in BV versus quasiconvexification in $W^{1,p}$: a model for the interaction between damage and fracture. *Calc. Var.*, to appear.

[40] Fonseca, I. and M. Katsoulakis. Minimizing movements and mean curvature evolution equations associated to a phase transitions problem. *J. Diff. and Int. Eq.*, to appear.

[41] Fonseca, I. and D. Kinderlehrer. Lower semicontinuity and relaxation of surface energy functionals. To appear.

[42] Fonseca, I., D. Kinderlehrer and P. Pedregal. Energy functionals depending on elastic strain and chemical composition. *Calc. Var.* **2** (1994), 283–313.

[43] Fonseca, I. and J. Malý. Relaxation of multiple integrals. To appear.

[44] Fonseca, I. and J. Malý. Relaxation of multiple integrals under constraints. In preparation.

[45] Fonseca, I. and P. Marcellini. Relaxation of multiple integrals in subcritical Sobolev spaces. *J. Geom. Anal.*, to appear.

[46] Fonseca, I. and S. Müller. Relaxation of quasiconvex functionals in $BV(\Omega; \mathbf{R}^p)$ for integrands $f(x, u, \nabla u)$. *Arch. Rat. Mech. Anal.* **123** (1993), 1–49.

[47] Francfort, G. A. and J. J. Marigo. Stable damage evolution in a brittle continuous medium. *Eur. J. Mech.*, A/Solids **12**, **2**, (1993), 149–189.

[48] Fusco, N. and J. E. Hutchinson. A direct proof for lower semicontinuity of polyconvex functionals. Preprint 1994.
[49] Gangbo, W. On the weak lower semicontinuity of energies with polyconvex integrands. *J Math. Pures et Appl.* **73, 5** (1994)
[50] Giaquinta, M., G. Modica and J. Souček. Cartesian currents and variational problems for mappings into spheres. *Preprint C.M.U.*, Pittsburgh 1992.
[51] Gdoutos, E. E. *Fracture Mechanics Criteria and Applications*, Kluwer, Dordrecht, 1990.
[52] Malý, J. Weak lower semicontinuity of polyconvex integrals. *Proc. Royal Soc. Edinburgh* **123A** (1993), 681–691.
[53] Malý, J. Weak lower semicontinuity of polyconvex and quasiconvex integrals. *Vortragsreihe*, Bonn, 1993.
[54] Malý, J. Lower semicontinuity of quasiconvex integrals. *Manuscripta Math.*, to appear.
[55] Marcellini, P. Approximation of quasiconvex functions and lower semicontinuity of multiple integrals quasiconvex integrals. *Manus. Math.* **51** 1985, 1–28.
[56] Marcellini, P. On the definition and the lower semicontinuity of certain quasiconvex integrals. *Ann. Inst. H. Poincaré, Analyse non Linéaire* **3** (1986), 391–409.
[57] Morrey, C. B. *Multiple Integrals in the Calculus of Variations*, Springer-Verlag, Berlin, 1966.
[58] Mumford, D. and J. Shah. Boundary detection by minimizing functionals. *Proc. IIIE Conf. on Computer Vision and Pattern Recognition*, San Francisco, 1985.
[59] Mumford, D. and J. Shah. Optimal approximationsby piecewise smooth functions and associated variational problems. *Comm. Pure and Appl. Math* **42** (1989), 577–685.
[60] Ziemer, W. P. *Weakly Differentiable Functions.* Springer-Verlag, Berlin, 1989.

Progress in Nonlinear Differential Equations
and Their Applications, Vol. 25

Magnetoelastic Interactions

DAVID KINDERLEHRER

Center for Nonlinear Analysis and Department of Mathematics
Carnegie Mellon University, Pittsburgh, PA 15213-3890, USA

ABSTRACT. To understand the role played by the complex microstructural arrangements in a giant magnetostrictive material, we were led to develop a micromagnetic theory where equilibria are described by the oscillatory statistics of approximating or minimizing sequences of a variational principle. The issues we encountered and the effectiveness of the theory are discussed in the context of experiment and simulation. Predictions based on the theory may be used to offer specific recommendation for improvement of actuator and sensor materials.

1. Introduction

Magnetoelastic interactions are ubiquitous: magnetizing a substance induces a strain, called magnetostriction[1]. That this strain may have preferential directions, and not be merely a uniform expansion or contraction, was first recognized by Joule, cf. [16] for additional historical comments. We witness in this way an interaction between electromagnetic and mechanical fields where symmetry plays an important role. The most common manifestation of the corresponding interaction between electric and mechanical fields is piezoelectricity. Most of our experiences with magnetostriction are adverse, like the vibration of a transformer in an electric substation and the heat consequently generated, which must be dissipated. At the other end of the scale, the production of magnetic storage devices, hard discs and reading/writing heads for example, involves choosing materials and designs to minimize undesirable strains. Exploitation of the Joule effect either for sensing or transducers, like sonar, began in the late 1920's but by the end of the war, magnetostrictive materials had been abandoned in favor of piezoelectric materials. The piezoelectric effect, the strain suffered by a material upon application of an electric field, was discovered by P. Curie in the final years of the last century. Remarkable piezoelectric materials, actually ferroelectrics, based on perovskite structures, began to be found during the war. Renewed interest in magnetostrictive materials owes to the discovery of high magnetostriction in $TbFe_2$ by A. E. Clark in 1971 and the subsequent discovery by him that the pseudobinary alloy of $Tb_xFe_{-x}Fe_2$ ≈ 0.3, had both high magnetostriction and low anisotropy [7]. This material, called Terfenol-D, has the property of giant magnetostriction. The short Table 1 offers a

[1]Supported by AFOSR 91 0301 and NSF DMS 911572 and by the ARO and the NSF through the Center for Nonlinear Analysis at Carnegie Mellon University. Computational resources provided by the NSF through a grant to the Pittsburgh Supercomputing Center.

Material	relative elongation in parts/million
PZT (piezoelectric)	100–300
Ni	40
Terfenol-D	1500–2000

Table 1. Comparison of relative length elongation

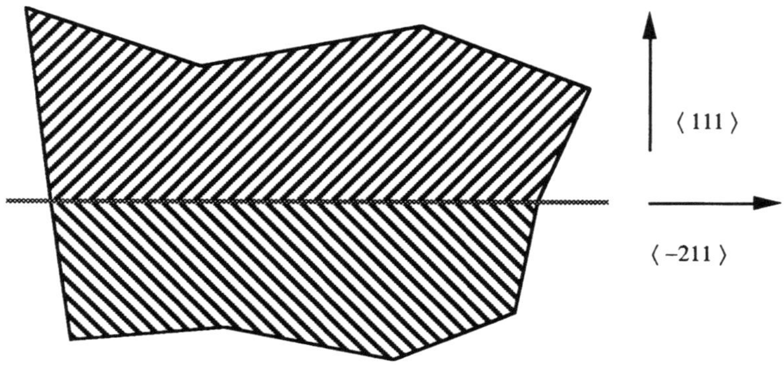

Figure 1: Schematic depiction of the microstructure in a sample of $TbDyFe_2$ illustrating the herringbone structure of two sets of lamellar fine structures. The shaded midrib line represents the position of a growth twin. Crystallographic directions are with reference to the high temperature nonmagnetic phase.

comparison. Piezoelectric materials are inexpensive but contain within them the seeds of their destruction, since an excessive field can initiate failure simply by the accumulation of space charge on repeated cycling (burning). Magnetic materials are quite expensive but do not have any such field related failure mechanism.

Terfenol-D condenses with a dendritic lamellar structure, cf. Figure 1, consisting of growth twins remaining fixed relative to the material during magnetostriction and separating lamellar structures [24], [26]. Understanding the nature of the magnetoelastic response requires identifying this microstructure and elucidating its role. The presence of several stable variants suggests that the material has a free energy with a multiple well structure, highly non convex, and thus is a nonlinear material. The theory we propose, based on our experience with martensitic materials, relies on determining the potential well structure precisely, [22], [23], [24], [25], [26]. It has its origins in the fundamental work on micromagnetics by W. F. Brown [4] and the elastic dielectric by R. A. Toupin [44] and incorporates the ideas about symmetry and phase transformations of J. L. Ericksen [17]. The spatially oscillatory fine structure indicates a competition between the free energy and other effects, leading to a certain frustration in the material, [21], [22]. A classical description of configurations may not be available even in unloaded equilibrium.

We have learned to understand this frustration, and systems where lower semicontinuity fails in general, through study of oscillatory statistics of the configurations, which is just another jargon for the Young measures associated to sequences which generate, in some suitable way, the equilibria [42], [45]. In effect, we are in the company of approximating sequences which converge weakly but never strongly. The advantage of the Young measure method is that through the minors relations, or the weak continuity of minors, and the variational principle, the oscillatory statistics are compelled to be consistent with energy considerations.

Moreover, the material exhibits an unusual magnetostriction curve which also would be useful to explain. In the course of a loading cycle, the process is not in its potential energy equilibrium wells, although it is probably nearby. The theory we have mentioned above will not apply exactly. One approach to this issue is to attempt a simulation of the loading cycle based on as simple as possible assumptions. Our efforts in this direction are described in §4. The experimental magnetostriction curve is hysteretic, namely, it traverses a path on increasing the applied magnetic field and another on decreasing the magnetic field, and thus a good simulation should display this property as well. The familiar magnetization vs. applied field picture universally found in textbooks is a paradigm of hysteresis. Generally accepted is the notion that defects and domain walls offer resistance to easy rotation of the magnetization and tend to result in metastable configurations which may be different on loading and unloading. This results in hysteresis but does not give any quantitative mechanism for its prediction: depending on the various defects in a given sample, a particular loading-unloading curve will be observed. A more fundamental explanation must come to terms with describing the notion of metastability as it is experienced by the material. We offer some views about this. Very similar phenomena are observed in other nonlinear processes, like the extension of a shape memory ribbon [36], [37], and may be quite dependent on the crystallographic orientation of the sample, Shield [40], and the shearing of a thin shape memory plate, cf. Ball, Chu, and James [3].

We are discussing here joint work with Richard James [24], [26] and Ling Ma [28], [30]. In addition we are touching on current investigations in progress with them and Antonio De Simone. Our theory was based in part on the attempt to explain the work of Lord and his collaborators [1], [2], [9], [10], [43] and some features of [8]. Very recently additional confirmation of our theory has come from the experiments of Dooley and DeGraef [15].

2. Formulation

In the constitutive theory we propose, the location of the potential wells has a prominent role while surface energies, magnetic domain wall energies, and similar effects are neglected. This is a large body approximation and is particularly appropriate owing to the highly mobile character of the lamellar structure, [11]. In this way, energy minimizing fine phase laminar twin systems and fine phase magnetic domain structures tend to limits of infinite fineness. Useful information about variant arrangement and location as well as macroscopic state functions, eg. energy and stress, is obtained by analysis of the Young measure extrema of the variational principle. The method is particularly useful in deciding where in the body fine structure will arise.

Introduce a stored energy density for the "single crystal", a portion of the material that derives from a single lattice, that depends on the deformation gradient $F \in \mathbb{M}$, 3×3 matrices, magnetization $m \in \mathbb{S}^2$, and temperature $\theta \in \mathbb{R}$,

$$W(F, m, \theta) \tag{2.1}$$

subject to the conditions of frame indifference

$$W(QF, mQ^T, \theta) = W(F, m, \theta), \qquad Q \in \mathrm{SO}(3), \tag{2.2}$$

and material symmetry

$$W(FP, m, \theta) = W(F, m, \theta), \qquad P \in \mathbb{P}, \tag{2.3}$$

where $\mathbb{P}$ is an appropriate crystallographic point group. Let $\Omega \subset \mathbb{R}^3$ denote a reference domain.

The point group $\mathbb{P}$ is determined by the symmetry of the non magnetic (untransformed) high temperature phase. We suppose that this stable phase is represented by the deformation gradient $F = \mathbb{1}$ and that its symmetry is given by the proper cubic (of order 24). This means that the reference domain Ω is taken to represent a homogenous body of cubic symmetry and that $W(\mathbb{1}, 0, \theta) = \min W(F, m, \theta)$ for high temperatures $\theta > \theta_{cr}$.

In its magnetic phase, where $\theta < \theta_{cr}$, the material suffers a distortion from the untransformed state and a magnetic moment appears. We then find a positive definite symmetric matrix U_1, the transformation strain, and a (non zero) vector m such that $W(U_1, m, \theta) = \min W(F, m, \theta)$. Local magnetic saturation requires that $|m| = f(\theta)$ is a function of temperature. Likewise, U_1 is a function of temperature for $\theta < \theta_{cr}$. We shall suppress this dependence and, in fact, normalize our energy so that $|m| = 1$ for $\theta < \theta_{cr}$.

The symmetry and frame indifference properties permit us to conjugate U_1 by SO(3) to obtain additional positive definite symmetric U_k and m_k, namely the set

$$\{(P^T U_1 P, mP) : P \in \mathbb{P}\} = \{(U_1, m_1), \ldots, (U_n, m_n)\}. \tag{2.4}$$

The complete collection of energy minima is then given by the wells

$$\Sigma = \bigcup_{1 \le k \le n} \mathrm{SO}(3)(U_k, m_k) = \bigcup_{1 \le k \le n} \{(QU_k, m_k, Q^T) : \ Q \in \mathrm{SO}(3)\} \tag{2.5}$$

and we assume that W assumes its minimum value exactly on Σ.

The next step is to understand our particular material and to write an energy for the growth twin crystal. In Terfenol-D, onset of ferromagnetism is associated with a stretch of the high temperature unit cell along a main diagonal parallel to the magnetization. W achieves its minimum energy on the eight pairs $(U_i, \pm m_i)$,

$$U_i = \eta_1 \mathbb{I} + (\eta_2 - \eta_1) m_i \oplus m_i, \qquad i = 1,2,3,4, \quad \text{and} \tag{2.6}$$

$$m_1 = \frac{1}{\sqrt{3}}(1,1,1), \qquad m_2 = \frac{1}{\sqrt{3}}(1,-1,-1),$$

$$m_3 = \frac{1}{\sqrt{3}}(-1,1,-1), \qquad m_4 = \frac{1}{\sqrt{3}}(-1,-1,1).$$

The typical configuration of TbDyFe2 rods consists of parallel growth twins. To model these, we assume that the entire rod is a composite for which we must introduce an inhomogeneous energy $W(F, m, x)$, $x \in \Omega$, cf. [24]. The lower lamellar structure arises from a rotation of 180° about the $m_1 = \frac{1}{\sqrt{3}}(1,1,1)$ axis of the original upper lattice. Denoting by R_o this rotation, the energy of the lower portion is given by $W(FR_o, m)$. Assuming coordinates to have been arranged so that the two regimes are separated by $x \cdot m_1 = 0$, we arrive at an energy density for the composite given by

$$W(F, m, x) = \begin{cases} W(F, m), & x \cdot m_1 > 0, \\ W(FR_o, m), & x \cdot m_1 < 0). \end{cases} \tag{2.7}$$

Note that R_o is not a symmetry operation of the original energy and, although holding invariant the wells of $(U_1, \pm m_1)$, gives a different set of wells with transformation strains and magnetizations

$$U_i' = \eta_1 \mathbb{I} + (\eta_2 - \eta_1) m_i' \otimes m_i', \ \pm m_i', \ \text{ with } \ m_i' = m_i R_o, \ i = 1,2,3,4. \tag{2.8}$$

From experiment, we know that $(\eta_2 - \eta_1)/\eta_2 \approx 2.4 \times 10^{-3}$, [1].

Maxwell's equations for magnetostatics may be introduced into this framework by

$$\mathrm{div}_y(\nabla_y v + \frac{1}{\det \nabla y} m) = 0 \text{ in } \mathbb{R}^3. \tag{2.9}$$

where v is a potential for the magnetic field and y is the deformation.

In this fashion we may write the energy of the virtual configuration $y = y(x)$, $m = m(x)$ in the mixed reference/spatial form (with temperature henceforth omitted)

$$E(y, m) = \int_\Omega W(\nabla y, m, x) dx + \frac{1}{2} \int_{\mathbb{R}^3} |\nabla_y v|^2 dy, \tag{2.10}$$

where the second term is the magnetic field energy, subject to the constraints,

$$\mathrm{div}_y(-\nabla_y v + \frac{1}{\det \nabla y} m) = 0 \text{ in } \mathbb{R}^3 \text{ and } |m| = 1 \text{ in } y(\Omega). \tag{2.11}$$

When the system is subject to an applied, assumed constant, magnetic field H, the energy of the virtual configuration $y = y(x)$, $m = m(x)$ is

$$E(y, m) = \int_\Omega \{W(\nabla y, m, x) - H \cdot m \det \nabla y\} dx) + \frac{1}{2} \int_{\mathbb{R}^3} |\nabla_y v|^2 dy). \tag{2.12}$$

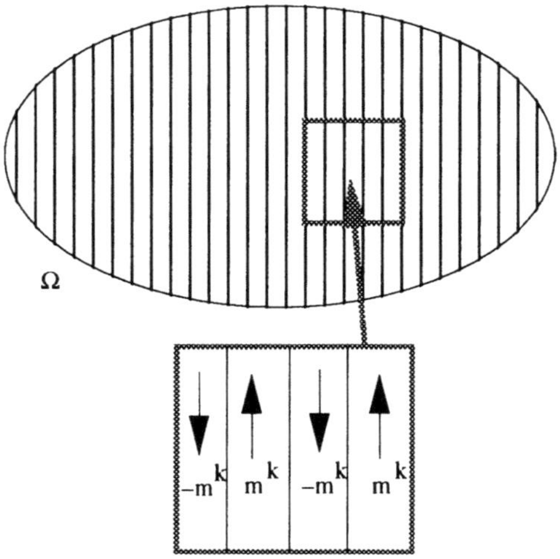

Figure 2: Depiction of a minimizing sequence of magnetizations. Note that $\operatorname{div} m^k = 0$ in Ω. The oscillations of the (m^k) are necessary to diminish the magnetic field energy.

From (2.7), we may also write the energy in the form

$$E(y,m) = \int_{\Omega} W(\nabla y, m, x)dx + \frac{1}{2}\int_{y(\Omega)} \frac{1}{\det \nabla y} m \cdot \nabla_y v \, dy. \tag{2.13}$$

If there is an $F_o \in \mathbb{M}$ and m_o, $|m_o| = 1$, with $W(F_o, m_o) = W(F_o, -m_o)$, then we may construct a minimizing sequence to show that

$$\inf E(y,m) = \min W|\Omega|. \tag{2.14}$$

However it need not be achieved and, in general, we do not expect it will be. Thus the representation of solutions naturally leads to the use of some averaging device, in our case Young measures and solutions of the minimization problem will be sought in this form. The importance of (2.11) is that it provided a criterion for energy minima. Its proof (2.11) involves showing that the magnetic field energy can be reduced to zero, which we accomplish by means of this fact [21], [29]:

If $f^k \in L^\infty(\mathbb{R}^n, \mathbb{R}^n)$ *satisfy*

$$\operatorname{div} f^k = 0 \textit{ in } H^{-1}(\mathbb{R}^n) \textit{ and } f^k \rightharpoonup 0 \textit{ in } L^2(\mathbb{R}^n),$$

then $m^k = \chi\Omega f^k$ *satisfy* $\operatorname{div} m^k \to 0$ *in* $H^{-1}_{\mathrm{loc}}(\mathbb{R}^n)$, *where* Ω *is a nice bounded open set.*

A minimizing sequence will have the appearance of Figure 2. The fact has been used in a creative way to construct the relaxation of (2.9) in the rigid case, [38].

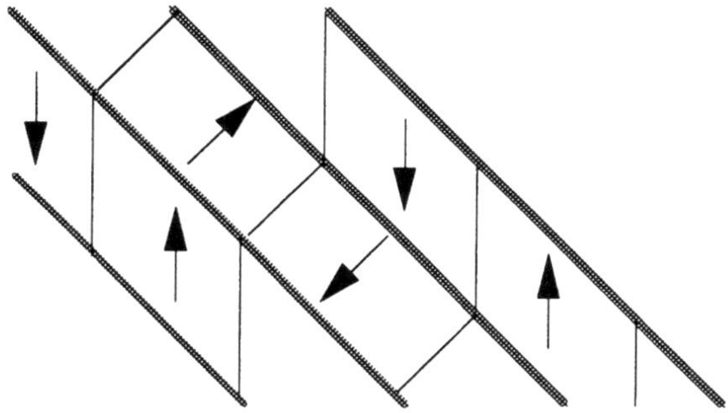

Figure 3: Depiction of a minimizing sequence of elastic domains, separated by shaded lines, and of magnetic domains.

3. Equilibrium microstructures and the mechanism of magnetostriction

To begin, we restrict attention to the upper laminate Ω^+ and seek lamellar solutions of (2.14). A simple lamellar structure should satisfy the mechanical twinning equation and its electromagnetic analogue, so as not to increase the magnetic field energy. Given energy wells $SO(3)(U,m)$ and $SO(3)(U^*,m^*)$, these relations are

$$RU^* = U(\mathbb{1} + a \otimes n) \text{ and } (m^*(RU^*)^{-T} - mU^{-T}) \cdot n = 0, \tag{3.1}$$

for some $R \in SO(3)$ and $a, n \in \mathbb{R}^3$. Note first that
(i) (3.1) is satisfied with $U^* = U$ and $m^* = -m$ and
(ii) if (3.1) is satisfied by (U,m) and (RU^*, m^*R^T), then it is satisfied by $(U,-m)$ and $(RU^*, -m^*R^T)$.

Now we observe that given any (U_i, m_i) and (U_j, m_j), we may satisfy (3.1) in two ways with $n \parallel m_i \pm m_j$ and $a \parallel m_i \mp m_j$. A minimizing sequence has the appearance of Figure 3. Each domain of mechanical distortion must display a magnetic substructure in order to diminish the magnetic field energy just as in Figure 2. A distinguishing feature of this type of analysis is that the actual scale of the magnetic domain substructure need not be finer than that of the mechanical laminate. This is confirmed in [15].

In this manner we find, for each pair (i,j), a Young measure supported on Σ of the form

$$\nu = \frac{1}{2}(1-\lambda)(\delta_{(U_i,m_i)} + \delta_{(U_i,-m_i)}) + \frac{1}{2}\lambda(\delta_{(RU_j,m_jR^T)} + \delta_{(RU_j,-m_jR^T)}), \tag{3.2}$$

with limit deformation and magnetization

$$F = (1-\lambda)U_i + \lambda RU_j \text{ and } m = 0. \tag{3.3}$$

where $0 \le \lambda \le 1$. It is not necessary that λ be constant, but let us assume this. There are two choices of rotation in (3.2) corresponding to the choices of normal n described above. Hence, qualitatively, there are twelve possible lamellar structures.

There are also twelve possible lamellar structures for the lower configuration Ω^-, leading to a possible 144 possiblilties for laminates, not accounting for different

$\left(\frac{1\,2}{1'2'}\right), \left(\frac{3\,4}{3'4'}\right)$ $\left(\frac{1\,2}{1'2'}\right)_r, \left(\frac{1\,3}{1'3'}\right), \left(\frac{1\,4}{1'4'}\right), \left(\frac{1\,3}{1'3'}\right)_r, \left(\frac{1\,4}{1'4'}\right)_r$ $\left(\frac{2\,3}{2'3'}\right)_r, \left(\frac{2\,4}{2'4'}\right)_r$ $\left(\frac{3\,4}{3'4'}\right)_r$

$\left(\frac{2\,3}{2'3'}\right), \left(\frac{2\,4}{2'4'}\right)$

Figure 4: Appearance of the predicted microstructures on the $(0-11)$ plane. For $(\frac{34}{3'4'})_r$, the lamella are parallel to the $(0-11)$ plane.

phase fractions. However the lower laminate with limit deformation F' must be related to the upper one F by

$$F' - F = b \otimes m_1 \tag{3.4}$$

in order for coherence to be valid across the growth twin interface. After arduous calculation, we found that (3.4) could be satisfied only if the lower and upper laminates are related by the rotation R_o, and the phase fractions above and below are equal. In other words, given (3.3), F' has the form

$$F' = (1-\lambda)QU_i' + \lambda QR'U_j', \text{ with } m_i' = m_i R_o \text{ and } m_j' = m_j R_o. \tag{3.5}$$

Hence there are only twelve possibilites. We use the terminology $(\frac{ij}{i'j'})$ or $(\frac{ij}{i'j'})_r$ to denote the arrangement of (3.3)–(3.5) composed of the twinned or reciprocally twinned variants U_i and U_j in Ω^+ and U_i' and U_j' in Ω^-.

We can now seek to explain the giant magnetostrictive effect by an analogy with twinning in martensitic materials. When a magnetic field H is applied parallel to the (-211) direction, the magnetization vectors in the lamella rotate towards H. The system decreases its energy by choosing the mechanical variant most agreeable to the magnetization. Thus the applied field induces an exchange of stability among elastic domains creating a much larger elongation than would be available by mere stretching of the material lattice.

The issue which now arises is whether or not all the possibilities $(\frac{ij}{i'j'})$ or $(\frac{ij}{i'j'})_r$ have the same magnetostrictive properties. Aside from their different geometric appearances, do they even posess the same kinematic and mechanical properties? The answer is no. A complete discussion would take us beyond the scope of this report, but, roughly speaking, any of the combinations $(\frac{ij}{i'j'})$ for $i, j = 2, 3, 4$, the individual lamella are themselves compatible across the growth twin boundary. That is, instead of the fine phase coherence described in (3.4), the exact coherence condition

$$QU_i') - U_i = bi \otimes m_1 \text{ and } QR'U_j' - RU_j = b_j \otimes m_1 \tag{3.6}$$

is valid, cf. [27] for a simple discussion. This probably permits a more flexible structure, for example, the phase fraction can vary along the growth twin while it cannot for a four gradient fine phase structure, but it does not provide the maximum magnetostriction, [26]. This is a property of the combination $(\frac{1\,2}{1'2'})$. To improve the magnetostrictive response of the material, we would like to assist the formation of this structure, for example, by attempting to grow the sample with as planar as possible a growth twin boundary and applying a compressive stress along the (-211) axis in order to exploit the special properties of this variant structure. Both of these techniques are known to the American manufacturer of Terfenol-D.

4. Simulation

Simulations of magnetic and magnetostrictive behavior based on micromagnetic theory, as described above, exhibit hysteresis. Their highly nonlinear character involves both short range anisotropy and elastic fields and dispersive demagnetization fields. Hysteresis occurs even in the absence of an imposed dynamical mechanism, for example, a Landau-Lifshitz-Gilbert dissipative equation for the magnetic moment, and is symptomatic of the way the system navigates a path through local minima of its energy space. It is not sensitive to the particular method: we implement continuation based on the conjugate gradient method, although the same results were obtained by other methods (eg., Newton's) as well. We strive to attain an efficient algorithm with careful attention dedicated to the treatment of the demagnetization energy. It is robust: computational experiments confirm that the shape of the loop is invariant over several decades of mesh refinement. Nonetheless, the propensity of optimization procedures to become marooned at local extrema when applied to nonconvex situations presents a fundamental challenge to analysis. Understanding and controlling such phenomena present the opportunity to develop predictive tools and diagnostics.

The full three dimensional nonlinear theory is beyond our grasp at this time. We have proposed a rigid model and a linear elastic model in two dimensions which perform well, but not to our satisfaction. The computations for rigid ferromagnets have been reviewed in [28], so we briefly describe the linear magnetoelastic material here, without indulging in an analysis of it.

A stored energy density in three dimensions for a cubic material with (111) easy magnetization axes is given by

$$\varphi(\varepsilon, m) = \varphi_{\mathrm{el}}(\varepsilon) + \varphi_{\mathrm{el/mag}}(\varepsilon, m) + \varphi_{\mathrm{an}}(m), \quad \varepsilon = \text{ linear strain},$$

where

$$\begin{aligned}
\varphi_{\mathrm{el}}(\varepsilon) &= \text{ typical cubic linear elasticity density} \\
\varphi_{\mathrm{el/mag}}(\varepsilon, m) &= b_1 \Sigma \varepsilon_{ii} m_i^2 + b_2(\varepsilon_{12} m_1 m_2 + \varepsilon_{23} m_2 m_3 + \varepsilon_{31} m_3 m_1) \\
\varphi_{\mathrm{an}}(m) &= K \Sigma m_i^4, \quad K > 0.
\end{aligned}$$

We project this energy onto an $(0-11)$ plane and refer to the (-211) direction as e_1 and the (111) direction as e_2. Retaining ε as the two dimensional (plane) strain and relabelling the magnetization a, we obtain a two dimensional energy density.

In our first trials, we have assumed the elastic portion to be isotropic. This gives rise a new density

$$\varphi(\varepsilon,\alpha) = \mu\Sigma\varepsilon_{ij}^2 + \frac{1}{2}\lambda(\operatorname{tr}\varepsilon)^2 - b\,g(\varepsilon)f(\alpha) - b'\Sigma\varepsilon_{ij}\alpha_i\alpha_j + \kappa\,f(\alpha)^2,$$
$$g(\varepsilon) = \frac{1}{3}\varepsilon_{11} - \frac{2}{3}\sqrt{2}\varepsilon_{12}, \quad f(\alpha) = \left(\frac{1}{3}\alpha_1 - \frac{2}{3}\sqrt{2}\alpha_2\right)\alpha_1, \tag{4.1}$$

where μ and λ are Lamé constants and b, b', and κ are all positive.

The simulation consists in computing by continuation in the field parameter H the energy

$$E(H,\alpha) = \int_\Omega (\varphi(\varepsilon,\alpha) - H\alpha_1)dx + \int_{\mathbb{R}^2} |\nabla u|^2 dx,$$

subject to

$$\Delta u = \operatorname{div}\alpha\chi_\Omega \text{ in } \mathbb{R}^2 \text{ and } |\alpha| = 1 \text{ in } \Omega.$$

The domain Ω is a rectangle with one axis parallel to the direction of the applied field, i.e., it represents a cross section of the Terfenol rod.

A second type of simulation corresponds to the growth twin laminate configuration. Here the domain Ω is divided horizontally into two rectangles Ω^+ and Ω^- and a stored energy φ is determined by taking (4.1) in Ω^+ and replacing the factor $-\frac{2}{3}\sqrt{2}$ by $+\frac{2}{3}\sqrt{2}$ in (4.1) for Ω^-. The two curves are slightly different. The magnetostriction curves do bear a close resemblance to the experimental magnetostriction curve as reported in [8], but we are not satisfied with our ability to develop a fine structure in the computation nor with our ability to determine the effect of fine phase coherent structures in the laminate. We are pursuing these questions.

References

[1] Al-Jiboory, M. and Lord, D.G., Study of the magnetostrictive distortion in single crystal terfenol-D by x-ray diffraction, IEEE Trans Mag 26, 2583–2585.

[2] Al-Jiboory, M. and Lord, D.G., Bi, Y.J., Abell, J. S., and Hwang, A.M.H., and Teter, J.P., Magnetic domains and microstructural defects in Terfenol-D, J. Appl. Phys. (to appear).

[3] Ball, J. M., Chu, C., and James, R. D., Metastability and hysteresis in elastic crystals (to appear).

[4] Brown, W.F., 1966, Magnetoelastic Interactions, Vol. 9 of Springer Tracts in Natural Philosophy (ed. C. Truesdell), Springer-Verlag.

[5] Bruno, O., Leo, P., and Shields, T., Transient heat transfer effects on the pseudoelastic behavior of shape-memory wires, Acta Met. Mat., 5, 191–201.

[6] Chipot, M., Kinderlehrer, D., and Ma, L., to appear.

[7] Clark, A. E., 1980, Magnetostrictive rare earth — Fe_2 compounds, Ferromagnetic Materials, Vol 1 (Wohlfarth, E. P. ed) North Holland, 532–589.

[8] Clark, A. E., 1992, High power rare earth magnetostrictive materials, Recent Advances in Adaptive and Sensory Materials, (Rogers, C.A.. and Rogers, R.C., eds), Technomic, 387–397.

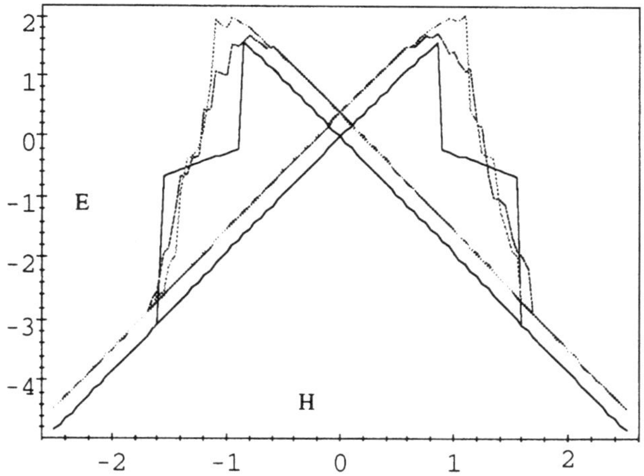

Figure 5: Computed energy curves for two dimensional model of Terfenol-D. The lowest curve is without induced field energy, the middle curve is the single crystal, and the upper curve is the growth twin laminate. The shoulders in the lowest curve, and the consequent gradual descents of energy in the upper two, are features of the material. The precise nature of the shoulders can be predicted from (4.1).

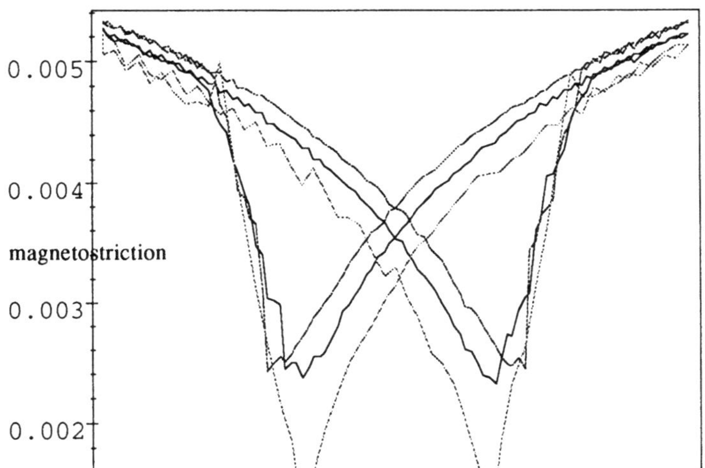

Figure 6: Segments of the computed magnetostriction curves for two dimensional model of Terfenol-D. The lowest curve is without induced field energy, the middle curve is the single crystal, and the upper curve is the growth twin laminate.

[9] Clark, A. E., Teter, J. P., and Mc Masters, O. D., 1988, Magnetostriction 'jumps' in twinned $Tb_{0.3}Dy_{0.7}Fe_{1.9}$, J. Appl. Phys., 63, 3910–3912.

[10] Clark, A. E., Verhoven, J. D., Mc Masters, O. D., and Gibson, E. D., 1986, Magnetostriction of twinned [112] crystals of Tb.27Dy.73Fe2, IEEE Trans Mag. 22, 973–975.

[11] De Simone, A., The effect of applied loads on the magnetostrictivie response of a terfenol-D-type material: a micromagnetic analysis, (to appear).

[12] De Simone, A., 1993, Energy minimizers for large ferromagnetic bodies, Arch. Rat. Mech. Anal., 125, 99–144.

[13] De Simone, A., 1994, Magnetization and magnetostrictive curves from micromagnetics, J. Appl. Phys., 76, 7018–7020.

[14] De Simone, A., 1994, Macroscopic response of magnetostrictive materials to applied magnetic fields and loads, Proc. SPIE vol. 2192, (Banks, H.T., ed.), 2–11.

[15] Dooley, J. and De Graef, M., TEM study of twinning and magnetic domains in Terfenol D (to appear).

[16] Ericksen, J.L., 1987, Twinning of crystals I, Metastability and Incompletely Posed Problems, IMA Vol. Math. Appl. 3, (Antman, S., Ericksen, J.L., Kinderlehrer, D., Müller, I., eds) Springer, 77–96.

[17] Ericksen, J.L., 1991, On kinematic conditions of compatibility, J. Elas. 26, 65–74.

[18] Giles, R., et al., 1992, Micromagnetics of thin film cobalt-based media for magnetic recording, Comp. in Phys., 6, 53–70.

[19] James, R. D. and Kinderlehrer, D., 1990, An example of frustration in a ferromagnetic material, Nematics: Mathematical and Physical Aspects, (Coron, J.-M., Ghidaglia, J.-M., and Hélein, F., eds), Kluwer NATO ASI series, 201–222.

[20] James, R. D. and Kinderlehrer, D., 1990, Frustration in ferromagnetic materials, Cont. Mech. Therm., 2, 215–239.

[21] James, R. D. and Kinderlehrer, D., 1992, Frustration and microstructure: an example in magnetostriction, Prog Part Diff Eqns: calculus of variations, applications (Bandle, C. et. al., eds) Pitman Res Notes Math 267, 59–81.

[22] James, R. D. and Kinderlehrer, D., 1992, Twinned structures in terfenol, Proc. ICOMAT-92, Wayman, M. and Perkins, J., eds), 269–274.

[23] James, R. D. and Kinderlehrer, D., 1993, A theory of magnetostriction with application to $TbDyFe_2$ Phil. Mag. B, 68, 237–274.

[24] James, R. D. and Kinderlehrer, D., 1993, Mathematical approaches to the study of smart materials, Mathematics in Smart Structures, Smart Structures and Materials '93, Proceedings SPIE vol. 1919, (Banks, H. T., ed.) 2–18.

[25] James, R. D. and Kinderlehrer, D., Laminate structures in martensite (to appear).

[26] James, R. D. and Kinderlehrer, D., 1994, Theory of magnetostriction with application to Terfenol-D, J. Appl. Phys., 76, 7012–7014.

[27] Kinderlehrer, D., and Ma, L., 1994, Computational hysteresis in modeling magnetic systems, IEEE Trans. Mag., 30.6, 4380–4382.

[28] Kinderlehrer, D., and Ma, L., 1994, Simulation of hysteresis in nonlinear systems, Math. and Control in Smart Structures, Proc. SPIE vol. 2192, (Banks, H.T., ed.), 78–87.

[29] Kinderlehrer, D., and Ma, L., The hysteretic event in the computation of magnetism and magnetostriction, Proc. Nonlinear Diff. Eqns. and their Appl., College de France Sem, Brezis, H. and Lions, J.-L., eds., (to appear).

[30] Kinderlehrer, D., and Ma, L., The hysteretic event in the computation of magnetization (to appear).

[31] Luskin, M. and Ma, L., 1992, Analysis of the finite element approximation of microstructure in micromagnetics, SIAM J. Num Anal., 29, 320–331.

[32] Luskin, M. and Ma, L., 1993, Numerical optimization of the micromagnetics energy, Mathematics in Smart Structures, Smart Structures and Materials '93, Proceedings SPIE vol. 1919, (Banks, H. T., ed.) 19–29.

[33] Ma, L., 1993, Computation of magnetostrictive materials, Mathematics in Smart Structures, Smart Structures and Materials '93, Proceedings SPIE vol. 1919, (Banks, H. T., ed.) 47–54.

[34] Ma, L. and Walkington, N., On algorithms for nonconvex optimization, to appear.

[35] Müller, I. and Xu, H., 1991, On the pseudoelastic hysteresis, Acta Metall., 39, 263–271.

[36] Ortin, J., 1992, Preisach modeling of hysteresis for a pseudoelastic Cu-Zn-Al single crystal, J. Appl. Phys., 71, 1454–1461.

[37] Pedregal, P., 1994, Relaxation in ferromagnetism: the rigid case, J. Non. Sci., 4, 105–125.

[38] Rogers, R. C., 1991, A nonlocal model for the exchange energy in ferromagnetic materials, J. Int. Eqns. Appl., 3, 85–127.

[39] Shield, T., Orientation dependence of the pseudoelastic behavior of single crystals of Cu-Al-Ni in tension (to appear).

[40] Stoner, E. C. and Wohlfarth, E. P., 1948, A mechanism of magnetic hysteresis in heterogeneous alloys, Phil. Trans. Royal Soc. (London) Sect. A, 240, 599–642.

[41] Tartar, L., 1979, Compensated compactness and applications to partial differential equations, Nonlinear analysis and mechanics: Heriot Watt Symposium, Vol I V (Knops, R., ed.) Pitman Res. Notes in Math. 39, 136–212.

[42] Teter, J.P., Mahoney, K., Al-Jiboory, M., Lord, D., and McMasters, O.D., 1991, Domain observation and magnetostriction in TbDyFe twinned single crystals, J. Appl. Phys., 69, 5768–5770.

[43] Toupin, R., 1956, The elastic dielectric, J. Rat. Mech. Anal., 5, 849–915.

Contributors

LUIGI AMBROSIO, Istituto di Matematiche Applicate, Università, Via Bonanno 25/B, 56126 Pisa, Italia

GABRIELE ANZELLOTTI, Dipartimento di Matematica, Università, 38050 Povo, Trento, Italia

ALEXIS BONNET, Université de Cergy-Pontoise 8, le Campus, 95033 Cergy-Pontoise Cx, France

ITALO CAPUZZO DOLCETTA Dipartimento di Matematica, Università La Sapienza, P.le Aldo Moro 2, 00185 Roma, Italia

MICHELE CARRIERO, Dipartimento di Matematica, Università, Via Arnesano, C.P. 193, 73100 Lecce, Italia

VICENT CASELLES, Departamento de Matematica i Informàtica, Università Illes Balears, 07071 Palma, Balears, España

BARTOMEU COLL, Departamento de Matematica i Informàtica, Università Illes Balears, 07071 Palma, Balears, España

BERNARD DACOROGNA, Départment des Mathematiques, Ecole Politecnique Fédérale, 1015 Lausanne, Switzerland

GIANNI DAL MASO, S.I.S.S.A., Via Beiruth 4, 34013 Trieste, Italia

ENNIO DE GIORGI, Scuola Normale Superiore, Piazza dei Cavalieri 7, 56126 Pisa

PEDRO MARTINS GIRÃO, Instituto Superior Tecnico, Av. Rovisco Pais, Lisboa, Portugal

IRENE FONSECA, Department of Mathematics, Carnegie Mellon University, Pittsburgh, PA 15213, USA

JEAN-PIERRE HAEBERLY, Department of Mathematics, Fordham University, Bronx, NY 10458, USA

DAVID KINDERLEHRER, Center for Nonlinear Analysis and Department of Mathematics, Carnegie Mellon University, Pittsburgh, PA 15213, USA

ROBERT V. KOHN, Courant Institute, New York University, New York, NY 10012, USA

ANTONIO LEACI, Dipartimento di Matematica, Università, Via Arnesano, C.P. 193, 73100 Lecce, Italia

PAOLO MARCELLINI, Dipartimento di Matematica, Università, V.le Morgagni 67/A, 50134 Firenze, Italia

MARIO MIRANDA, Dipartimento di Matematica, Università, 38050 Povo, Trento, Italia

JEAN-MICHEL MOREL, Ceremade, Université Paris Dauphine,
Pl. de Lattre de Tassigny, 75775 Paris Cx 16, France

UMBERTO MOSCO, Dipartimento di Matematica, Università La Sapienza,
P.le Aldo Moro 2, 00185 Roma, Italia

LINO NOTARANTONIO, Dipartimento di Matematica, Università La Sapienza,
P.le Aldo Moro 2, 00185 Roma, Italia

JAYANT SHAH, Mathematics Department, Northeastern University,
Boston, Mass. 02115, USA

HALIL METE SONER, Department of Mathematics, Carnegie Mellon University,
Pittsburgh, PA 15213, USA

ITALO TAMANINI, Dipartimento di Matematica, Università,
38050 Povo, Trento, Italia

FRANCO TOMARELLI, Dipartimento di Matematica, Politecnico di Milano,
Piazza Leonardo da Vinci 32, 20133, Milano, Italia

List of participants

Emilio Acerbi (Parma)
Micol Amar (Pavia)
Luigi Ambrosio (Pisa)
Gabriele Anzellotti (Trento)
Valeria Bacchelli (Milano)
John M.Ball (Edinbourgh)
Giovanni Bellettini (Pisa)
Marco Biroli (Milano)
Cristiana Bondioli (Pavia)
Alexis Bonnet (Cergy-Pontoise)
Guy Bouchitté (Toulon)
Primo Brandi (Perugia)
Italo Capuzzo Dolcetta (Roma)
Michele Carriero (Lecce)
Maria Cristina Cerutti (Milano)
Antonin Chambolle (Paris)
Antonio Cianci (Milano)
Claudio Citrini (Milano)
Antonio Corbo Esposito (Cassino)
Alessandra Coscia (Parma)
Bernard Dacorogna (Lausanne)
Florangela Dal Fabbro (Milano)
Gianni Dal Maso (Trieste)
Riccardo De Arcangelis (Salerno)
Giuseppe De Cecco (Lecce)
Virginia De Cicco (Roma)
Ennio De Giorgi (Pisa)
Annelise Defranceschi (Parma)
Francesco Doveri (Milano)
Loris Faina (Perugia)
Stefano Finzi Vita (Roma)
Irene Fonseca (Pittsburgh)
Marco Fuhrman (Milano)
Fabio Gastaldi (Brescia)
Filippo Gazzola (Alessandria)
Ugo Gianazza (Pavia)
Gianni Gilardi (Pavia)
Federico Girosi (Boston)
Maurizio Grasselli (Milano)
Richard D. James (Minneapolis)
David Kinderlehrer (Pittsburgh)
Robert V. Kohn (New York)
Maria Rosaria Lancia (Roma)
Antonio Leaci (Lecce)
Francesco Leonetti (L'Aquila)
Paola Loreti (Roma)
Daniela Lupo (Milano)
Enrico Magenes (Pavia)
Paolo Marcellini (Firenze)
Riccardo March (Roma)
Clelia Marchionna (Milano)
Antonella Martinolli (Milano)
Giacomo Mazzini (Pisa)
Mario Miranda (Trento)
Giuseppe Modica (Firenze)
Jean-Michel Morel (Paris)
Stefano Mortola (Milano)
Umberto Mosco (Roma)
Vincenzo Nesi (L'Aquila)
Tullia Norando (Milano)
Vladimir Oliker (Atlanta)
Pirro Oppezzi (Genova)
Diego Pallara (Lecce)
Giuliana Palmieri (Bari)
Maurizio Paolini (Pavia)
Giovanni Prouse (Milano)
Luisa Rossi Costa (Milano)
Anna Salvadori (Perugia)
Giuseppe Savaré (Pavia)
Klaus Schilling (Bonn)
Raul Serapioni (Milano)
Francesco Serra Cassano (Trento)
Jayant Shah (Boston)
Antonio Siconolfi (Roma)
Italo Tamanini (Trento)
Nicoletta Tchou (Roma)
Franco Tomarelli (Milano)
Vincenzo M. Tortorelli (Pisa)
Carla Vaghi (Milano)
Amina Vasconi (Milano)
Claudio Verdi (Milano)
Vincenzo Vespri (Pavia)
Enrico Vitali (Parma)
Anna Zaretti (Milano)

Progress in Nonlinear Differential Equations and Their Applications

Editor

Haim Brezis, Département de Mathématiques, Université P. et M. Curie 4, Place Jussieu, 75252 Paris Cedex 05, France *and*
Department of Mathematics, Rutgers University
New Brunswick, NJ 08903, U.S.A.

Progress in Nonlinear and Differential Equations and Their Applications *is a book series that lies at the interface of pure and applied mathematics. Many differential equations are motivated by problems arising in such diversified fields as Mechanics, Physics, Differential Geometry, Engineering, Control Theory, Biology, and Economics. This series is open to both the theoretical and applied aspects, hopefully stimulating a fruitful interaction between the two sides. It will publish monographs, polished notes arising from lectures and seminars, graduate level texts, and proceedings of focused and refereed conferences.*
We encourage preparation of manuscripts in some form of TeX for delivery in camera-ready copy, which leads to rapid publication, or in electronic form for interfacing with laser printers or typesetters.

Proposals should be sent directly to the editor or to: Birkhäuser Boston, 675 Massachusetts Avenue, Cambridge, MA 02139.

PNLDE 1+2 *Partial Differential Equations and the Calculus of Variations. Essays in Honor of Ennio De Giorgi*
F. Colombini / A. Marino / L. Modica / S. Spagnolo (Eds)

PNLDE 3 *Propagation and Interaction of Singularities in Nonlinear Hyperbolic Problems*
M. Beals

PNLDE 4 *Variational Methods*
H. Berestycki, I. Ekeland, J.-M. Coron (Eds)

PNLDE 5 *Composite Media and Homogenization Theory*
G. Dal Maso, G. Dell' Antonio (Eds)

PNLDE 6 *Infinite Dimensional Morse Theory and Multiple Solution Problems*
K.-C. Chang

PNLDE 7 *Nonlinear Diffusion Equations and Their Equilibrium States, 3*
Proceedings from a Conference held August 20-29, 1989 in Gregynog, Wales
N.G. Lloyd, W.M. Ni, L.A. Peletier, J. Serrin (Eds)

PNLDE 8 *An Introduction to G-Convergence*
G. Dal Maso

PNLDE 9 *Differential Inclusions in Nonsmooth Mechanical Problems. Shocks and Dry Friction*
M.D.P. Monteiro Marques

PLNDE 10 *Periodic Solutions of Singular Lagrangian Systems*
Antonio Ambrossetti and Vittorio Coti Zelati

PNLDE 11 *Nonlinear Waves and Weak Turbulence with Applications in Oceanography and Condensed Matter*
N. Fitzmaurice, D. Gurarie, F. McCaughan and W.A. Woyczynski (Eds)

PNLDE 12 *Seminar on Dynamical Systems*
Euler International Mathematical Institute, St. Petersburg, 1991
S. Kuksin

PNLDE 13 *Ginzburg-Landau Vortices*
F. Bethuel, H. Brezis, F. Hélein

PNLDE 14 *Variational Methods in Image Segmentation*
J.-M. Morel, S. Solimini

PNLDE 15 *Topological Nonlinear Analysis: Degree, Singularity, and Variations*
M. Matzeu, A. Vignoli

PNLDE 16 *Analytic Semigroups and Optimal Regularity in Parabolic Problems*
A. Lunardi

PNLDE 17 *Blowup for Nonlinear Hyperbolic Equations*
S. Alinhac

PNLDE 18 *The Heat Kernel Lefschetz Fixed Point Formula for the Spin-c Dirac Operator*
J.J. Duistermaat

PNLDE 19 *Nonlinear Dynamical Systems and Chaos*
H.W. Broer, I. Hoveijn, F. Takens, S. van Gils

PNLDE 20 *Topics in Geometry.*
Honoring the Memory of Joseph D'Atri
S. Gindikin

PNLDE 21 *Partial Differential Equations and Mathematical Physics. The Danish-Swedish Analysis Seminar, 1995*
L. Hörmander, A. Melin

PNLDE 22 *Partial Differential Equations and Functional Analysis. In Memory of Pierre Grisvard*
J. Cea, D. Chenais, G. Geymonat, J.L. Lions (Eds)

PNLDE 23 *Geometry of Harmonic Maps*
Y. Xin

PNLDE 24 *Minimax Theorems*
M. Willem

PNLDE 25 *Variational Methods for Discontinuous Structures. Applications to image segmatition, continuum mechanics, homogenization Villa Olmo, Como, 8-10 September 1994*
R. Serapioni, F. Tomarelli (Eds)

MIX
Papier aus verantwortungsvollen Quellen
Paper from responsible sources
FSC® C105338

If you have any concerns about our products,
you can contact us on
ProductSafety@springernature.com

In case Publisher is established outside the EU,
the EU authorized representative is:
Springer Nature Customer Service Center GmbH
Europaplatz 3, 69115 Heidelberg, Germany

Printed by Libri Plureos GmbH
in Hamburg, Germany